NUREG-1922

Computational Fluid Dynamics Analysis of Natural Circulation Flows in a Pressurized-Water Reactor Loop under Severe Accident Conditions

Manuscript Completed: February 2010
Date Published: March 2010

Prepared by
C.F. Boyd and K.W. Armstrong

Office of Nuclear Regulatory Research

ABSTRACT

Computational fluid dynamics is used to predict the natural circulation flows between a simplified reactor vessel and the steam generator of a pressurized-water reactor (PWR) during a severe accident scenario. The results extend earlier predictions of steam generator inlet plenum mixing with the inclusion of the entire natural circulation loop between the reactor vessel upper plenum and the steam generator. Tube leakage and mass flow into the pressurizer surge line are also considered. The predictions are utilized as a numerical experiment to improve the basis for simplified models applied in one-dimensional system codes that are used during the prediction of severe accident natural circulation flows. An updated inlet plenum mixing model is proposed that accounts for mixing in the hot leg too. The new model is consistent with the predicted behavior and accounts for flow into a side mounted surge line if present. A density-based Froude number correlation is utilized to provide a method for determining the flow rate from the vessel to the hot leg directly from the conditions at the ends of the hot leg pipe. This provides a physically based approach for establishing the hot leg flows. The mixing parameters and correlations are proposed as a best-estimate approach for estimating the flow rates and mixing in one-dimensional system codes applied to severe accident natural circulation conditions. Sensitivity studies demonstrate the applicability of the approach over a range of conditions. The predictions are most sensitive to changes in the steam generator secondary side temperatures or heat transfer rates to the steam generator. Grid independence is demonstrated through comparisons with previous models and by increasing the number of cells in the model. A further modeling improvement is suggested regarding the application of thermal entrance effects in the hot leg and surge line. This work supports the U.S. Nuclear Regulatory Commission studies of steam generator tube integrity under severe accident conditions.

FOREWORD

Steam generator tubes comprise a majority of the reactor coolant system pressure boundary and therefore their integrity is important to ensuring safe operation of the plants. During November 2000, after the Indian Point 2 steam generator leakage event, NRC developed a Steam Generator Action Plan (SGAP) to consolidate NRC activities related to steam generators and to ensure that issues are appropriately tracked and completed. In May 2001, NRC revised the SGAP to address a differing professional opinion on steam generator tube integrity. This plan is just one facet of NRC's overall strategy to maintain safe operation of nuclear power plants, increase public confidence in agency regulatory actions, and ensure efficient use of NRC resources.

One aspect of the SGAP is the analysis of the risk of low probability severe accident-induced tube failures. Tube failure during a severe accident has the potential for radioactive release to the environment that bypasses containment. An important component of this analysis is the thermal-hydraulic prediction of the reactor coolant system response during severe accident conditions. The focus is on the failure timing for reactor coolant system (RCS) components as well as the impact of safety systems and operator actions on tube integrity. Improving the prediction of the thermal-hydraulic conditions that challenge RCS boundaries during a severe accident is the subject of this report.

Thermal-hydraulic analysis of the RCS provides the temperature and pressure conditions that challenge RCS components including the steam generator tubes. Temperature of the steam in the reactor coolant loops is influenced significantly by three-dimensional mixing behavior. System codes such as SCDAP/RELAP5 do not implicitly model this mixing and rely on predetermined mixing and flow parameters to account for this behavior. One principal source of information is a set of experiments at 1/7th scale (Westinghouse Electric Corporation, Research Project 2177-05). The Office of Nuclear Regulatory Research has been using computational fluid dynamics (CFD) techniques to extend the experimental results for different steam generator geometries and conditions.

This report describes the completion of a detailed analysis of reactor coolant loop natural circulation flows under severe accident conditions in a specific Westinghouse pressurized-water reactor geometry. The predictions are used like a numerical experiment to determine flow and mixing parameters for tuning system codes such as SCDAP/RELAP5 or MELCOR. Earlier studies at 1/7th scale (NUREG-1781) demonstrate the applicability of the technique, and the initial extension of the methods to full-scale conditions (NUREG-1788) demonstrated the benefits of using CFD in extending the experimental results. This current study improves upon the earlier work with updated modeling techniques and the incorporation of review comments and lessons learned. The updated analyses provides an improved mixing model formulation along with best-estimate flow and mixing parameters for application to system code models of similar Westinghouse plants. This effort reduces modeling uncertainties and improves the technical basis of the NRC staff understanding of the severe accident system behavior.

TABLE OF CONTENTS

Figures

Tables

EXECUTIVE SUMMARY

This report summarizes an updated analysis of severe accident natural circulation flows between the reactor vessel upper plenum and the steam generator using computational fluid dynamics (CFD). The analysis supports system code modeling of severe accident induced tube ruptures completed as part of the U.S. Nuclear Regulatory Commission's (NRC's) steam generator action plan (SGAP). The updated analysis builds upon earlier validation work at 1/7th scale (NUREG-1781) and the initial extension of this approach to full-scale conditions (NUREG-1788). Review comments and lessons learned from these prior studies are incorporated into this updated analysis to improve the technical basis of the staffs understanding of reactor coolant system behavior under severe accident conditions. Predictions and recommendations from this analysis support NRC's latest SCDAP/RELAP5 system modeling (NUREG/CR-6995) of severe accident-induced tube failures as part of section 3.4 (thermal-hydraulic work) in the SGAP.

NRC implemented the SGAP to confirm the robustness of risk-informed licensing decisions, reduce modeling uncertainties, and improve the technical basis for future licensing actions. One objective of the plan is to investigate the time-dependent thermal-hydraulic conditions in the hot leg, surge line, and steam generator under severe accident conditions. This report is the culmination of several detailed studies that incorporated a state-of-the-art CFD model to predict the three-dimensional natural circulation flows of interest. A set of data at 1/7th scale provides a fundamental understanding of the natural circulation flows, and these data were used to benchmark the CFD predictions of inlet plenum mixing in an initial study (NUREG-1781). The test data are limited to a specific steam generator geometry and do not cover all of the geometries or conditions of interest as pointed out in a follow-on study of full-scale steam generators (NUREG-1788). The current study builds upon the work in NUREG-1788 by incorporating a model of the entire natural circulation loop from the reactor vessel upper plenum to an improved model of the steam generator tube bundle. The numerical predictions are used in the manner of a numerical experiment to extend and enhance our understanding of the three-dimensional severe accident flows. A set of modeling coefficients are determined from the predictions to improve the basis for the models used in one-dimensional system codes to simulate the natural circulation flows and mixing.

The CFD model utilized represents the primary side of a steam generator, the hot leg, a portion of the pressurizer surge line, and a simplified reactor vessel upper plenum along with a small portion of a vessel. The FLUENT 6.3 CFD code is used for the analysis. The predictions qualitatively show all of the flow features observed experimentally in the hot leg and steam generator regions. The natural circulation flows are unsteady in nature. To obtain average values, a transient simulation is completed using fixed boundary conditions to produce a series of predictions that are combined to obtain the average behavior. Average mass flows and temperatures are predicted throughout the flow domain and used to find the coefficients for mixing and flow models that are suitable for use in one-dimensional system code modeling.

This study used the predictions to find a discharge coefficient related to a density-based Froude number correlation. This correlation is proposed as a means to determine the hot leg mass flows in a system code model. The approach relates the hot leg flow rate to a density-based Froude number. The proportionality constant is the discharge coefficient, and it is found to be 0.12 for this reactor geometry. The discharge coefficient did not show significant sensitivities to the modeling parameters, and this physically based approach is easily adaptable to system code models. Establishing the hot leg mass flows in a system code model in this manner addresses a past criticism by the Advisory Committee on Reactor Safeguards (ACRS) regarding the scaling of the heat distribution to the steam generators that relied on the results of tests at 1/7th scale.

This study expands the inlet plenum mixing model to include the hot leg mixing and entrainment. This updated approach is more consistent with the CFD predictions and results in a higher mixing fraction and a more realistic estimate of the recirculation ratio. The mixing fraction and recirculation ratio determined from these predictions are 0.96 and 2.4, respectively. This new mixing model can also account for the presence of a side mounted surge line. Specific recommendations are outlined regarding the use of these mixing parameters in system code models.

Since system code predictions have shown that the reactor loop with the pressurizer can have the earliest tube failures under some conditions, it is important to consider the impact of the flows into the pressurizer surge line. The mass flow into the surge line is accounted for in the updated mixing model. In addition, the CFD predictions are used to define the mixture of flows that enter the surge line during periods of countercurrent flow. For side-mounted surge lines, the flow entering the surge line enters equally (50:50) from the upper and lower hot leg flows. For a top-mounted surge line, it is determined that about 75 percent of the flow comes from the flow in the upper hot leg. The models are appropriate for loop flows that do not involve a pressurizer surge line simply by setting the surge line mass flow to zero. The top-mounted surge line is not fully investigated in this study.

The most important aspect of these predictions is the determination of the range of tube temperatures in the tube bundle and specifically, the hottest temperature range. System code models typically use a single representative tube for the hot tube flows, and the temperature is a mass-averaged value for the entire group of tubes carrying the hot flow. With over 1,000 tubes carrying hot flow in a prototypical steam generator, a significant variation in temperature can exist between the highest and lowest temperature tube in the hot tube region. A normalized temperature is defined to make the results easy to apply under a variety of conditions. A value of 1.0 represents the temperature of the flow from the vessel upper plenum, and a value of 0.0 represents the temperature of the flow returning to the inlet plenum through the cold flow tubes. Tube entrance temperatures will always fall between 0 and 1 on this scale. The average normalized temperature of the hottest tube in the bundle for the base case is found to be 0.43, and peak temperatures reaching the tubes intermittently approach 0.625. This study recommends that an average normalized temperature value of 0.5 be used for system code evaluations to bound the observed low-frequency variations in the average tube inlet temperature.

A series of sensitivity studies are completed to establish the stability of the predictions and to quantify the sensitivity of the results to variations in some of the key modeling parameters. Specific input parameters are systematically varied to determine the impact on the predicted mixing and flow parameters of interest. The secondary side temperature is found to be the most significant parameter that was varied. This is consistent with the earlier studies (NUREG-1788) that showed the significance of the heat transfer rate to the secondary side. Equal magnitude variations in the vessel temperature showed some of the same characteristics. One important finding from the sensitivity studies is the lack of significant change in the discharge coefficient. This parameter changed by less than 10 percent over all of the studies. The relatively constant nature of this parameter confirms that this is a good method for establishing the hot leg flow rates in system code models over a wide range of conditions. A qualitative study on tube leakage rates demonstrated that leakage rates below 1.5 kg/s do not have an effect on the overall flows and mixing. The countercurrent flow pattern continued in the presence of tube leakage rates up to 6.0 kg/s. At a leakage rate of 12 kg/s, very little mixing is predicted and the hottest tube temperatures approach the hot leg flow temperature. Under the assumed conditions, a leakage rate of 6 kg/s is similar to the leakage expected from a leak area equal to about 1 tube cross section.

On a qualitative basis, this study notes that the hot leg and surge line convective heat-transfer rates are underpredicted by system code models that rely on fully developed heat-transfer correlations in their pipe flow models. To determine a best-estimate heat transfer coefficient in these areas, the impact of the entrance region should be accounted for. Some data are provided for this purpose. Incorporating these effects into a system code model can double the convective heat-transfer coefficient in the entrance region of a hot leg and surge line. It is noted that severe accident induced failure points of the hot leg and surge line are anticipated to be near the initial entrance region of the pipes where the temperatures and heat transfer coefficients are highest.

The updated predictions outlined in this report build upon previous studies and provides an updated set of parameters for use in one-dimensional system codes to predict three-dimensional natural circulation flows in a Westinghouse pressurized-water reactor loop under severe accident conditions. The results are specific to the geometry and conditions utilized in this study and should not be applied universally. The same techniques, however, could be applied to study reactor systems under different conditions with alternate piping and inlet plenum designs.

1 INTRODUCTION

The U.S. Nuclear Regulatory Commission (NRC) has implemented a steam generator action plan[1] to study steam generator tube integrity. This plan includes evaluating the risk of temperature-induced tube rupture during low probability severe reactor accidents, which requires thermal-hydraulic predictions of the thermal and pressure loads on various parts of the reactor coolant system. System codes are used to predict the plant behavior under a variety of conditions. In specific severe accident scenarios, the system pressure remains high and the source of water to the steam generators fails, resulting in a dry condition. In addition, if significant leakage occurs on the secondary side of the steam generators, then the pressure will drop in the secondary side and the mechanical (pressure) load on the tubes will increase. This unlikely series of events leads the plant into a state referred to as a high-dry-low condition, which refers to the high primary system pressure, the dry steam generators, and the low pressure on the secondary side.

Under high-dry-low conditions, the steam generators lose their ability to transfer heat and the primary system boils down ultimately leading to substantial core uncovery, core heat up, and generation of significantly superheated steam and hydrogen. When the water level drops below the hot legs, the steam coming off the core sets up a three-dimensional counter-current natural circulation flow pattern that transfers heat out into the reactor coolant loops. The loop seals remain filled with water, and full loop circulation is blocked. Figure 1 shows a simplified sketch of this natural circulation flow pattern. Failures in the hot leg nozzle, pressurizer surge line connection in the loop with the pressurizer, and steam generator tubes are possible during this period. Failure of a steam generator tube can lead to containment bypass during a severe accident.

The natural circulation flows, mixing, and heat transfer in the hot leg, steam generator inlet plenum, and steam generator tubes play a significant role in determining what primary system components will fail and the timing of these failures. Creep rupture failures are typically predicted in either the hot leg, pressurizer surge line (if present), or possibly a steam generator tube. It is the details of the natural circulation flow rates, mixing, and heat transfer that are required to accurately predict the relative timing of the potential failures. If a significant component fails prior to a steam generator tube (e.g., the hot leg or surge line), the system will depressurize into the containment and the threat to the tubes is eliminated. If a tube fails prior to the hot leg or surge line, a potential exists for radioactive release to the environment. This fact drives the need for the improved understanding of the plant behavior during these types of severe accident conditions. Background information on modeling and the risk of severe accident-induced tube ruptures can be found in NUREG/CR-6285[2] and NUREG-1570[3]. A recent study, which has been used extensively as a reference for this report, is documented in NUREG/CR-6995.[4]

The thermal-hydraulic modeling of high-dry-low severe accident scenarios is performed with lumped parameter system analysis codes such as SCDAP/RELAP5 (SR5) or MELCOR. System analysis codes are able to predict the behavior of the entire reactor coolant system over extended periods of time. These codes, however, do not explicitly model the three-dimensional natural circulation flow phenomena that are critical to the examination of this issue. In instances where three-dimensional behavior is important, system code models are adjusted to be consistent with the expected flow conditions that are determined from experimental correlations and/or suitable multidimensional code predictions. The countercurrent natural circulation flow rates, mixing, and heat transfer associated with the flow pattern described by Figure 1 provides a specific example of a case where the system code models need to be augmented with experimental correlations or multi-dimensional predictions.

Test data are available from a 1/7th scale facility that provide valuable information on steam generator inlet plenum mixing and the natural circulation flows. These data, however, are limited to a single non-prototypical inlet plenum design and there are concerns related to the scaling of the tube bundle secondary side heat transfer. Previous analyses by the NRC staff documented in NUREG-1781[5] demonstrated that computational fluid dynamics (CFD) predictions can adequately predict the inlet plenum mixing observed in the 1/7th scale tests. A set of follow-on analyses documented in NUREG-1788[6] studied full-scale steam generators under severe accident conditions to extend the experimental results to prototypical conditions. These results demonstrated that details of the steam generator design can impact the results. This highlights the benefits of using CFD modeling to extend the experimental results. The full-scale CFD predictions (reference 6) provide valuable insights into the inlet plenum mixing behavior under a variety of full-scale conditions. After a review of these predictions, the Advisory Committee for Reactor Safeguards (ACRS) recommended[7] extending the modeling to include a prediction of the full natural circulation flow path between the vessel upper plenum and the steam generator. The modeling outlined in this report addresses the ACRS concern with the addition of a model for the reactor vessel upper plenum. The updated modeling also incorporates other improvements including a significantly improved tube bundle model.

The FLUENT (version 6.3) CFD code is used to predict the natural circulation flows between the reactor vessel upper plenum and the steam generator. FLUENT is a commercially available, general-purpose CFD code capable of solving a wide variety of fluid flow and heat transfer problems. The code solves the Reynolds-averaged Navier-Stokes equations on a finite volume mesh. The Navier-Stokes equations represent the mass, momentum, and energy conservation equations for a continuous fluid. Reynolds averaging creates the need for turbulence modeling to account for the turbulent diffusion of momentum and energy. The FLUENT code provides several turbulence modeling options. A fully unstructured meshing capability allows the code to be applied to complex geometries. Commercial CFD codes such as FLUENT are widely used in many industries today and are commonly used to predict mixing phenomena.

Each of the steps in a CFD analysis can influence the predicted results and must be considered carefully. The basic steps involve describing the physical model; defining the model domain, boundary conditions, and models; validating the solution; and completing sensitivity studies to quantify the impact of parameter assumptions. When considering the results of CFD predictions, the user should also consider the assumptions and limitations of the process. Further details on the fundamentals of CFD are found in the introductory text by Anderson[8], and reports are available that describe best practices including a recent CFD best practice guide focused on nuclear safety analysis[9] published by the Organization for Economic Cooperation and Development.

2 OVERVIEW OF ANALYSIS

The modeling outlined in this report is an extension of the analysis completed in NUREG-1788 for the model 44 and 50 steam generator designs. Where possible, the modeling assumptions are consistent with the previous work to maintain consistency and a connection to the validation work at 1/7th scale (NUREG-1781). Solver settings, turbulence options, and general node density remain the same. The major difference in the new model is the addition of a simplified representation of a reactor vessel upper plenum, the inclusion of a pressurizer surge line, and a refinement of the tube bundle model. These additions make the model more realistic and address a few of the limitations of the prior work. The addition of a reactor vessel component removes the need to specify mass flow as a boundary condition on the upper vessel end of the hot leg. The addition of a surge line allows the model to account for loss of mass from the hot leg to the surge line in some cases. Improvements in available computer capacity made it practical to model a larger number of tube flow paths and to accurately represent the tube bundle flow area. The improved tube bundle model required millions of extra computational cells. One other improvement in the current model is the addition of hydrogen. A species tracking model is included to track hydrogen through the loop and the material properties are representative of the steam-hydrogen mixture. In all aspects, the updated model is either consistent with or improves upon the previous modeling documented in NUREG-1788.

Transient simulations are not practical due to the relative size of the steam generator models and the extended time span of the transient. The model also does not include the solid components which would play a key role in any transient heat up analysis. A quasi-steady assumption is used and the boundary conditions are obtained from a snapshot of a SR5 prediction for a prototypical plant during a transient simulation. Boundary conditions are selected at a point in the scenario several minutes after the period of rapid core oxidation begins. This point represents a temperature range that is between the high and low temperature cases outlined in NUREG-1788. The conditions carried over from the system analysis include the temperature of the flow from the vessel upper plenum, the steam generator secondary side temperature, temperatures of the tube sheet structure over the inlet and outlet plenums, the mass flow from the hot leg into the pressurizer surge line, and a snapshot of the transient heat transfer rate from the tube flows to the tube walls. These heat transfer rates to the steam generator tube bundle walls are significant in that they establish a key component of the transient behavior in the quasi-steady CFD simulation. This transient heat transfer rate includes the heat transfer to the tube wall that heats up the tube structures as well as the heat that is transferred to the secondary side of the steam generator. The overall process of determining the boundary conditions is iterative since the CFD boundary conditions come from the SR5 predictions and the SR5 models are adjusted with coefficients from the CFD predictions. The conditions used in this study are the result of several iterations between the CFD and SR5 predictions.

The CFD predictions primarily support the modeling of the natural circulation flow rates, heat transfer, and mixing in the system code models. Updated models for the inlet plenum mixing and hottest tubes, the hot leg flow rates, surge line entrance flows, and the hot leg heat transfer coefficients are extracted from the CFD predictions or existing correlations. The updated models and coefficients are the principal result of this analysis effort.

2.1 Physical Domain Considered

Figure 2 shows the physical model domain used for these predictions. It consists of the primary side of a full-scale steam generator (similar to a model 44 or model 51) along with the hot leg, a portion of the pressurizer surge line, and a simplified portion of the vessel inside of the core barrel. The pressurizer loop is considered because SR5 predictions have indicated that the induced failures in this loop can occur earlier than failures in the other reactor loops. Non-pressurizer loops are considered by closing off the connection to the pressurizer surge line. The model is intended to simulate the three-dimensional natural circulation flows illustrated in Figure 1 which are present during periods of time when the pressurizer relief valve is closed. During these periods, pressure in the system is increasing which results in a small but significant flow from the hot leg to the pressurizer surge line. To model this behavior, a mass flow of steam and hydrogen (estimated from system code models) is allowed to exit the model through the surge line. The mass balance is maintained by an equivalent mass source at the bottom of the core region.

The tube bundle simulates a Westinghouse model 51 steam generator with 3,388 0.019685 m (0.775 inch) tubes with a tube pitch of 0.032537 m (1.281 inches). Modeling each tube individually is impractical, so the model uses tubes with a cross-sectional area equal to 9 prototypical tubes (model combines a 3 x 3 group of tubes into 1 tube). The tube pitch in the model is three times larger than the tube pitch in a model 51 steam generator. The innermost row of tubes in the model is blocked off to simulate a condition of 10- percent tube plugging which is a relatively common condition. The open tubes have a flow area equivalent to 90 percent of the flow area in a model 51 generator. The height of the tubes is setup to match the model 51 design. Figure 3 provides a cross-sectional view of the tube layout with the plugged tubes illustrated with an "X." Table 1 compares the specific details of the tube bundle model to a model 51 example. Modeling parameters, discussed later in the report, are utilized to ensure that the steam generator tube bundle is modeled with the appropriate heat transfer and flow resistance characteristics (see Appendix A).

Table 1. Steam Generator Model Dimensions

Tube Bundle Detail	CFD Model	Model 51 Generator
Number of tubes**	371	3388
Total flow area (includes plugged tubes)	1.01689 m^2	1.031131 m^2
Open flow area (10% plugging design point)	0.92854 m^2	0.9280 m^2
Tube pitch (square array)	0.09761	0.03254 m
Tube sheet thickness	0.5398 m	0.5342 m
Total height of tube bundle above tube sheet	10.624 m	10.592 m
Inlet plenum radius	1.5954 m	1.5954 m

** a total of 14 half tubes are counted as 7 tubes

The steam generator inlet plenum is a ½-sphere design with a radius of 1.5954 m (62.81 inches). The midplane of the sphere lies on the lower tube sheet face. The tube sheet is modeled as a boundary condition over the first 0.5398 m of the tube flow path.

The hot leg runs horizontally from the vessel upper plenum to the steam generator inlet plenum. The hot leg has a 50-degree upward bend at the inlet plenum and its axis is aligned 36.5 degrees from the normal to the divider plate that separates the inlet and outlet plenums. The steam generator design is not symmetric. The centerline distance of the hot leg from the edge of the vessel upper plenum to the entrance to the steam generator inlet plenum is 7.072 m. The horizontal section has a length of 5.784 m with the rest of the length taken up by the elbow that curves up into the inlet plenum. The diameter of the hot leg is 0.7366 m (29 inches).

Two pressurizer surge line connections are attached to the hot leg 2.889 m from the reactor vessel upper plenum. The base case and the majority of the predictions utilize the side-mounted surge line while the top-mounted surge line is blocked off keeping the inner wall of the hot leg smooth. For a sensitivity study, the side-mounted surge line can be blocked off and the upper surge line connection opened up. The surge line is a 0.2845 m (11.2 inch) pipe that connects to the center of the hot leg (either vertically or horizontally). Flows into the surge line exit at the end of the pipe. The pressurizer is not modeled. To simulate a non-pressurizer loop, both surge line connections are blocked off.

A simplified model for ¼ of the reactor vessel inside of the core barrel is included. The model extends from near the bottom of the active fuel to the upper dome. The core region includes three rings with the inner two rings representing the fuel region and the outermost ring representing the bypass region between the core and core barrel. Above the core are surfaces or regions representing an upper core plate, the upper plenum, the upper support plate, the upper head, and the upper dome. A porous media model is used to represent flow losses in these regions, and complete blockages are accomplished using interior walls where appropriate. The overall height of the components and the vessel radius is similar to a Westinghouse 4 loop reactor. The purpose of the vessel model is to ensure conditions in the upper plenum region that are consistent with those from the SR5 model predictions.

2.2 Base-Case Prediction

A single best-estimate prediction is made that utilizes boundary conditions derived from a SR5 prediction for a prototypical plant at a snap-shot in time when the core oxidation is starting to rise significantly. A separate analysis of the steam generator tubing is completed to ensure that the tube-bundle model yields the appropriate heat transfer and pressure drop characteristics (see Appendix A). This best-estimate prediction is referred to as the base-case prediction.

2.3 Sensitivity Studies

Starting with the best-estimate model developed for the base-case prediction, a series of sensitivity studies are completed to quantify the impact of changes in some of the key input parameters to the base-case model. These sensitivity studies included a consideration of vessel upper plenum temperatures, tube bundle heat loss and pressure drop, hydrogen mass fraction, surge line position and mass flow, turbulence modeling, and the addition of tube leakage. These sensitivity studies provide an improved understanding of the parameters that govern the natural circulation flow phenomena and the results help to quantify the uncertainty in the CFD and system-code predictions.

2.4 Modeling Inputs for System-Code Analysis

The CFD predictions are used to formulate updated models for the inlet plenum mixing, the hottest tube determination, surge line entrance flows, and the steam generator tube and hot leg flow rates. These models are developed specifically for use in a SR5 system code model during countercurrent natural circulation flows. Hot leg heat transfer was studied separately. Qualitative CFD predictions of hot leg heat transfer indicated higher heat transfer in the upper section of the hot leg nozzle compared to the original SR5 model. A review of this issue is completed, and values for improving the convective hot leg heat transfer modeling in the system code models are presented (see section 6).

3 CFD MODELING APPROACH

The physical geometry is represented with a finite volume mesh on which the governing Navier-Stokes equations are discretized and solved. All models are developed for the FLUENT version 6.3 unstructured CFD code.

The solutions are expected to be unsteady, so a transient solver is applied with steady boundary conditions. Predictions are obtained at fixed intervals and averaged to obtain an average solution for each case considered. Specific features of the CFD model and major assumptions are outlined below.

3.1 Basic Solver Settings

The basic solver options selected for this analysis are listed below.

- Transient Reynolds averaged Navier-Stokes solution with steady boundary conditions.
- 0.05 second time step.
- Vertical symmetry plane for ¼ upper plenum model.
- Reynolds stress turbulence model (2nd order) with non-equilibrium wall functions.
- Full buoyancy effects on turbulence (as defined in FLUENT model).
- Temperature-dependent thermal properties (steam and hydrogen) at constant pressure
- Hydrogen species tracking.
- Gravity.
- Segregated solver with 2nd-order differencing on momentum and energy.
- Loss coefficients added to vessel region and steam generator tubes.

3.2 Finite Volume Mesh

An effort is made to produce models with a high-quality mesh that represents the key features of the primary system flow path described in Section 2.1. To maintain a level of consistency between models, mesh quality and spacing are consistent with those used for NUREG-1788 in the hot leg region. The addition of a vessel component and a refined tube-bundle model increased the total number of computational cells to 7.8 million. Over 6 million of these cells are in the steam generator. In the bulk of the mesh, the cell aspect ratios are limited to 2 with the exception of the long, straight regions of the steam generator tubes where cell aspect ratios are stretched to 10 in the flow direction. Cell skew is minimized through careful grid spacing and with the use of hexagonal elements, where possible. Growth rates between neighboring cells are limited to 20 percent with few exceptions. The majority of the cells in the hot leg and plenums are of the same approximate dimensions.

Figures 4, 5, and 6 give an indication of the mesh design at a few key locations in the model. Figure 4 shows the lowest region of the tube-bundle mesh starting at the upper surface of the inlet plenum. The mesh is designed to have an average cell aspect ratio near 1.0 at the junction of the tube and the inlet plenum. The cells grow in the flow direction along the length of the tube and reach a maximum aspect ratio of 10. Figure 4 (right side) shows the mesh cross section for a few tubes. A total of 7 cells are located across the tube diameter, and the wall cell has a thickness of 0.006 m. Each tube has approximately 15,000 volumes between the inlet and outlet plenum.

Figure 5 shows the mesh on a plane aligned with the hot leg that passes through the inlet and outlet plenums. The mesh in the inlet plenum utilizes a hexagonal core design. The cells in the central region of the inlet plenum are 0.015 m cubes (aspect ratio = 1.0) aligned with the hot leg axis. A small transition region of tetrahedral elements exists between the central region and the wall, which is paved with 0.02 m triangles. The total number of cells in the inlet plenum is approximately one million. The high quality of the mesh in the bulk of the inlet plenum is ideal for the purposes of predicting the inlet plenum mixing and entrainment. The mesh in the outlet plenum is a coarse tetrahedral mesh. Mixing issues within the outlet plenum is not considered significant to this analysis. A grid sensitivity study is completed where the inlet plenum cell dimensions are cut in half and the number of cells in the inlet plenum is increased to eight million. This study indicated that the base mesh density is sufficient to predict the mixing.

Figure 6 shows the mesh in the hot leg on the hot leg centerline and at one cross section just before the elbow leading into the inlet plenum. A clearer view of the mesh cross section is provided on the right side of the figure. Mesh in the hot leg is designed to be consistent with the mesh used in reference 5. The wall cells are 0.015 m thick and a total of 30 cells are located across the diameter. Minimal stretching is employed for the majority of the hot leg resulting in aspect ratios near 1 for the bulk of the hot leg region. Cells are not stretched in this region because the flow in the hot leg is expected to be a complex countercurrent flow.

The mesh is the reactor vessel upper plenum maintains the cell quality from the hot leg. The mesh in the vessel component is mainly hexagonal with a small tetrahedral transition region.

3.3 Boundary Conditions

The boundary conditions establish a temperature difference between the reactor vessel upper plenum and the steam generator tube bundle that drives the natural circulation flows of interest. These conditions are established to match a snapshot in time of a transient SR5 simulation in order to setup conditions applicable to the severe accident conditions. The model establishes the hot temperatures in the vessel upper plenum by using a fixed temperature approach in the surrounding vessel regions. All flows in the fuel region and upper dome are set to a fixed temperature and this effectively controls the temperature in the upper plenum. These surrounding temperatures are adjusted until the flow leaving the upper plenum matches the predicted value from the SR5 model. The tube bundle model is setup to loose heat to the secondary side (boiler region) at a rate that is consistent with the transient SR5 predictions. This boundary condition allows the quasi-steady CFD prediction to have a heat transfer rate that includes the transient heating of the steam generator tube bundle. This makes the CFD predictions more applicable to transient analyses. The secondary side is set to a temperature consistent with the SR5 prediction. In addition, boundary conditions are established to provide a specified mass flow into the pressurizer surge line. This mass flow is replaced in the model by injecting mass flow through the bottom of the vessel component to simulate a steam-hydrogen mixture rising up from the core. The vessel component represents only ¼ of the full vessel within the core barrel, and symmetry planes are used on the vertical end faces of this section.

With the exception of the steam generator tubes and the fixed temperatures in the vessel component, all walls are assumed to be adiabatic and have a no-slip shear condition. Shear stress is computed using wall functions within the FLUENT code. Figure 7 gives an overview of the principal regions of the model. Details for specific regions are outlined below.

3.3.1 Tube Bundle

A tube-modeling approach was successfully used in previous work at 1/7th- and full-scale conditions (references 5 and 6, respectively). Limitations of this approach, as noted in the previous work, included an enlarged bundle flow area. The updated tube-bundle model, although still simplified, uses considerably more computational cells and utilizes the correct bundle flow area. Figure 3 shows a cross section of the updated tube bundle just above the tube sheet, and Table 1 provides some details on the tube-bundle geometry.

Boundary conditions for the tube walls are applied uniformly over a few key regions. These include the two halves of the tube sheet (inlet and outlet side) and the tubes above the tube sheet that are exposed to the secondary side (boiler region). It is assumed that the secondary side conditions (in the boiler region) are uniform and apply the same thermal conditions (temperature and heat-transfer coefficient) to all tube surfaces above the tube sheet. Although the assumption of uniform heat transfer over such a large region is a significant simplification, it is a practical necessity due to the lack of detailed flow-field information on the secondary side of the tube bundle. The system-code model utilizes only a few computational cells to cover the entire secondary side region around the tube bundle and temperatures from these regions are used to derive the secondary side conditions for the CFD model. The tube sheet is a relatively large mass with a high thermal inertia. The SR5 model studied for this report utilized separate heat structures for the inlet and outlet side of the steam generator. Inner wall heat structure temperature conditions from the SR5 prediction are used as fixed inner-wall temperature conditions for the steam generator tube walls in the tube-sheet region. Different temperatures are used for the inlet and outlet side tube sheets to reflect the fact that the tube sheet is predicted to be hotter on the inlet side of the steam generator.

The simplified tube-bundle model used for this analysis must be augmented to perform like a full-scale tube bundle with thousands of tubes. Although the thermal boundary conditions are setup to match the SR5 predictions for the sequence of interest, the tube bundle does not provide enough heat transfer surface area and the heat transfer must be augmented. The same analogy is true for the pressure drop. The hydraulic diameter of the tubes in the simplified tube bundle model is too large. Flow loss coefficients must be added to the model in the region of the tubes to increase the pressure drop and therefore match the expected behavior. Heat transfer rates are augmented by increasing the thermal conductivity of the fluid in the tube region. Both of these adjustments are applied using the porous media option in the FLUENT solver. Appendix A outlines the results and the process for determining the required loss coefficients and thermal conductivity values.

The base-case thermal boundary condition for the tubes in the tube sheet region is a fixed-wall temperature of 880K on the inlet side and 803K on the outlet side. The base-case boundary condition on the tube walls above the tube sheet is a convective condition on the exterior of the wall with a heat transfer coefficient of 86 W/m^2-K and an ambient temperature of 813K. The wall is assumed to be stainless steel with a thickness of 0.00127m. The convective conditions

9

are established to match the SR5 predictions of tube bundle heat loss. As noted earlier, the heat transfer is also augmented by increasing the thermal conductivity of the fluid in the tubes. Appendix A provides details of the tube bundle modeling approach.

3.3.2 Pressurizer Surge Line Mass Flow

In the SR5 predictions, the most severe challenge to the tubes occurs in the pressurizer loop under the highest pressure conditions. The CFD predictions are focused on the pressurizer loop for this reason and consider the time during the scenario when the pressurizer relief valve is closed and the pressure is rising towards the relief valve set point. It is during these periods of time that the counter-current natural circulation flows are established. During this time, the SR5 predictions indicate a mass flow from the hot leg to the pressurizer through the pressurizer surge line. This mass flow is the result of the pressure increasing in the system. The ultimate source of this mass flow is assumed to be the core region where steam (and hydrogen) is generated. To model this behavior, an inlet is set up on the bottom of the vessel model so that a mixture of steam and hydrogen can be injected into the core region at a given temperature. This injected mass ultimately leaves the system through an outlet at the end of one of the active pressurizer surge lines. The inlet mass flow is adjusted to obtain the desired mass flow from the hot leg to the pressurizer surge line. For the base-case model, 1.3 kg/s of a steam-hydrogen mixture (hydrogen mass fraction of 0.0055) at 1,250K enters the model in a vertical direction through the inlet at the base of the vessel region. This boundary condition results in an average mass loss to the pressurizer of 1.3 kg/s which is consistent with the SR5 prediction. The base-case model uses the side-mounted surge line and blocks off the top-mounted surge line. The other non-pressurizer loops are considered by shutting off the surge line connections completely.

3.3.3 Hot Leg, Surge Lines, and Plenums

The hot leg, pressurizer surge line, and steam generator plenum volumes are modeled directly with geometry that is representative of a prototype plant. A standard no-slip boundary condition is applied, and the wall roughness is assumed to be zero. A wall-function approach is used to calculate the shear stress. All of these walls are assumed to be adiabatic.

3.3.4 Upper Plenum and Vessel

The vessel component is setup to supply a hot steam mixture into the reactor vessel upper plenum at a temperature established to match the upper plenum temperatures from specific SR5 predictions. The vessel upper plenum attaches to the hot leg and has three annular rings designed to crudely simulate the reactor upper plenum. Figure 7 illustrates the vessel model and upper plenum. The outer ring of the upper plenum region is sized to represent the open space between the reactor internals and the hot leg junction. The inner two rings use a porous media approach with loss coefficients to simulate the blockages resulting from the reactor internals. This three-ring design extends from the base of the core region up through the upper head (see Figure 7). No heat transfer occurs in the upper plenum region, which serves as a mixing volume for the hot flows coming out of the core and the cold flows returning from the steam generator through the lower hot leg. A thin region representing the upper support plate is located above the upper plenum, and above this plate is the upper head, another supporting region, and the upper dome. A plate region representing the upper core plate is located below the upper plenum, and the core is located below this region. The elevations and radial extent of the vessel components approximate the design of a typical Westinghouse four-loop pressurized-water reactor vessel.

The core region was initially setup as a heat-generating porous body to simulate the energy generation and flow resistance of a core region during a severe accident. This approach added complexity to the model and solution convergence was not practical. A significant amount of trial and error adjustments were required to obtain the desired conditions in the upper plenum region. Flows in the entire system were unsteady by nature, and this added to the complexity. The approach was simply not practical with the available computer resources, and further simplifications were sought.

The vessel component objective changed to simply establish the desired severe accident conditions in the upper plenum region of the vessel. It is this region that interacts with the hot leg and steam generator loop and ultimately drives the natural circulation flows. Fixed temperature regions are established above and below the upper plenum to ensure that the flows coming into the upper plenum are at the desired temperature. Mixing and buoyancy-driven flows are not adjusted in the upper plenum region, and significant temperature variations are observed in this region. By adjusting the gas temperatures above and below the upper plenum region, the conditions in the upper plenum could be held relatively steady and this reduced the overall flow variations and made it practical to determine average loop flow behavior for a given set of conditions in the upper plenum. Although flows are predicted throughout the vessel region, they are not considered realistic due to the elimination of temperature differences (fixing of temperature) above and below the upper plenum.

As noted earlier, a mass flow inlet is established at the base of the core region on the inner ring. This inlet provided steam and hydrogen at a mass flow rate equal to the expected surge line mass flows (predicted with SR5) and at a temperature equal to the fixed temperature of the core region.

3.4 Material Properties

The working fluid is a steam-hydrogen mixture that is assumed to be at 2,400 psia. Fluid properties are adapted from several sources including the materials data base from RELAP5. The quasi-steady solutions represent a single point in the accident sequence, and pressure is assumed to remain constant. The properties are assumed to vary only with temperature. Specific data covering the range of expected temperatures are input to the FLUENT code in tabular form for both steam and hydrogen. The code uses linear interpolation (based on temperature) to find the thermal properties for each gas at a given temperature and uses mixing laws to determine the properties of the mixture. Density is obtained from a volume-weighted mixing law. The specific heat is determined as a mass fraction weighted average. Both the viscosity and thermal conductivity of the gas mixture are obtained from a mass-weighted mixing law. The mass diffusivity between the gases is based on a constant dilute approximation with a mass diffusivity of 2.9×10^{-5} m^2/s. This simplification is good for the well-mixed constant conditions used for this analysis. The constant mass diffusivity value (the laminar component) input to the model is rarely of consequence in problems of this sort where the turbulent diffusivity dominates any diffusion. The turbulent Schmidt number is left at the default value of 0.7. The hydrogen mass fraction assumed for the base case prediction is 0.0055.

Table 2 provides individual gas material properties. The density is provided as a function of temperature for both steam and hydrogen. Temperature-dependent steam properties also are provided for the specific heat, viscosity, and thermal conductivity. Constant properties for the specific heat (15,000 J/kg-K), viscosity (1.2e-5 W/m-K), and thermal conductivity (0.3 W/m-K) are used for the hydrogen.

Table 2. Material Properties

Density (kg/m³)			specific heat (J/kg-K)		viscosity (kg/m-s)		k – conductivity (W/m-K)	
T (K)	Steam	H$_2$	T (K)	Steam	T (K)	Steam	T (K)	Steam
700	65.98	5.588	700	3,864.0	700	0.0848	700	2.8e-5
800	50.85	4.890	800	2,867.4	800	0.0889	1,200	4.69e-5
900	42.84	4.346	900	2,613.7	900	0.1	2,500	9.92e-5
1,000	37.46	3.911	1,000	2,548.2	1,100	0.1236		
1,100	33.47	3.556	1,100	2,539.4	1,200	0.1528		
1,200	30.35	3.260	1,200	2,558.8	1,400	0.177		
1,300	-	3.009	1,400	2,634.6	2,000	0.210		
1,400	25.71	2.794	2,000	2,868.1				
1,600	22.36	2.445						

3.5 Solution Initialization, Monitoring, and Convergence

The predictions obtained indicated an unsteady buoyant plume in the inlet plenum of the steam generator and an unsteady wavy interface between the hot and cold flows in the hot leg. A transient solver is used for the predictions, and average values are computed over multiple realizations of the predicted flows to obtain a representative average value for the desired parameters of interest. To initiate the solutions, the velocity is set to zero everywhere, and temperatures are set by region. The vessel temperature is set to the core inlet temperature (1,250 K for the base case). Temperature in the steam generator and plenums is set to the tube secondary side temperature from the heat-transfer boundary condition (813 K for the base case). Finally, the remaining portions of the model between the vessel and steam generator (hot leg and surge lines) are set to the average of the hot and cold temperatures (1,031.5 K for the base case). With this initial condition, the solution quickly establishes the countercurrent flow pattern of interest. Gradually, the solution reaches a steady (on average) oscillating condition. This is monitored while the time accurate solver is executed for many hundreds of seconds. Key parameters, such as selected mass flows or temperatures, are plotted at every time step (0.05 s for base-case model) and monitored for convergence. In addition, the solution residuals are monitored at each time step to ensure the solution is converged. Finally, an overall mass and energy balance are checked. Once the solution appears to be converged, a series of full data sets are saved on 2-second intervals, and key results from these data are plotted. The longest oscillations are associated with the mass flow rate in the tube bundle.

One suggested measure of the solution period is the tube-bundle flow residence time. Approximating the average path length through the steam generator (inlet plenum to outlet plenum) at 22 m and using an average velocity in the tube bundle of 0.4 to 0.5 m/s results in a tube residence time in the range of 44 to 55 seconds. This is consistent with some of the larger observed solution oscillations. For the base-case prediction, a total of 140 full data sets are considered to get an idea of the degree of variations in the solution. The complete set of data, representing snapshots covering 280 seconds of solution time, is used to obtain average results for the base-case model. These same data are broken down into seven 40-second intervals (20 full data sets each) to estimate the solution variations. Sensitivity studies are completed using two or more sets of 40-second intervals (40 or more data sets).

3.6 Data Reduction

The solution procedure is stopped when a sufficient number of complete data sets have been saved. Each data set contains a record of all of the solution variables at each of the nearly 8 million computational cells. These data are reduced (data reduction) to a manageable set of engineering values of interest to facilitate the analysis. Each data file is read into the solver, and a script is utilized to extract key parameters of interest such as the hot leg or tube bundle mass flows and temperatures. These data are put into a spread sheet for further analysis. If the data do not show a long-term trend, the averaged results are considered a meaningful prediction of the system behavior. The predictions are then utilized as a numerical experiment to develop and define the system code modeling parameters.

3.7 Summary of Assumptions and Limitations

The completed CFD predictions give valuable insights into the three-dimensional natural circulation flows between the upper plenum of the reactor vessel and the steam generator. When considering the significance of these predictions, it is important to consider how they could be influenced by the modeling assumptions and limitations. To enhance the understanding of the predicted results, some of the major assumptions and limitations are highlighted below.

3.7.1 Tube-Bundle Model

The tube-bundle model is designed to match the pressure drop and heat-transfer characteristics of a full-scale steam generator tube bundle. The goal is to be able to predict the tube-bundle mass flows by correctly balancing the buoyancy driving force with the total pressure drop in the bundle. This is completed with a simplified tube-bundle model that uses 371 individual tubes that each have a cross-sectional area equivalent to 9 prototypical (0.785-inch ID) tubes. Because the tubes are larger than the prototype, the shear force needs to be augmented with a body force term in an attempt to match the pressure-drop characteristics over a range of velocities. Similarly, the thermal conductivity of the tube fluid is increased in an attempt to match the heat-transfer rates. Appendix A provides details of the tube-bundle model.

One obvious limitation of the tube-bundle model is the reduced number of tubes. Thousands of tubes are grouped and modeled with hundreds of individual flow paths. This simplification limits the resolution of tube-to-tube variations across the tube bundle. The hottest region of the plume occupies a region with a diameter equal to 5 tube spaces and therefore the peak of the plume is resolved with 5 data points. Another issue relates to tubes that reverse flow direction during the simulation. This occurs at the boundary between the hot and cold tube regions. It is suggested that it may be more difficult for the models tubes to change flow due to the larger size (increased inertia) per tube.

Another significant limitation of the tube-bundle modeling approach relates to the heat-transfer boundary condition. A uniform condition is applied over all of the tubes in the boiler region. Sensitivity studies demonstrate that the tube-bundle heat transfer affects the tube-bundle mass flows and other parameters important to the natural circulation flow rates. The assumption of uniform heat-transfer across large regions of the tube bundle could be significant. This analysis has not considered secondary side conditions with spatial variations in the tube-wall boundary conditions.

3.7.2 Grid Independent Solution

A few indicators are used to conclude that the solution is grid independent and that the mesh is adequate for this particular study. First, the mesh size in the inlet plenum is consistent with the mesh size used in the benchmarking exercise at 1/7th (NUREG-1781, reference 5). This previous work demonstrated that the mesh used was suitable for predicting the component-scale mass flows and average temperatures of interest. In this current work, the inlet plenum mesh quality is improved though the use of flow aligned cubic cells in the majority of the inlet plenum. The inlet plenum mixing, as judged by the range of temperatures entering the tube bundle, indicates no significant grid dependency.

To further investigate the possibility of grid dependence, the inlet plenum mesh density is increased from around one million to eight million cells. The majority of the eight million cells were flow aligned cubic volumes. No significant change in the inlet plenum mixing is observed in the refined mesh case. Peak temperatures entering the bundle changed by less than 0.5 percent and overall mass flows and ratios remained the same. It is concluded that the base case mesh is adequate for predicting the loop flows and mixing.

3.7.3 Adiabatic Walls

The overall CFD model sets a high-temperature condition in the core region, and the only heat sink is in the tube bundle. The hot leg and inlet plenum walls are assumed adiabatic. A solid–wall, heat-transfer model is not practical in a steady-state analysis since the solid walls would simply heat up to the fluid temperature. Specifying a fixed three-dimensional heating rate for the steady-state analysis also is difficult because the flow field is unsteady. The approach used greatly simplifies the complexity of the thermal boundary condition on the hot leg and inlet plenum and is considered reasonable in light of the small surface area of the hot leg and inlet plenum in comparison to the steam generator tube bundle.

3.7.4 Vessel Model

The vessel model uses a highly simplified design to represent some of the major regions of the vessel. The upper plenum region, which interacts with the loop flows directly, is created from three annular rings. The outermost ring has no obstructions and the inner two rings are modeled with a porous media approach to simulate the flow losses associated with the structures in the upper plenum. All of the assumptions and limitations of the porous media approach apply to these regions. A fixed temperature condition above and below the upper

14

plenum was applied to control the temperatures in the upper plenum. These conditions change the overall buoyancy-driven flows within the vessel component. In addition, the vessel component is made up of regions that are modeled using a porous media approach in a crude attempt to simulate the flow resistances of the vessel components. Two symmetry planes also are used for the vessel component to create a 90-degree sector of the vessel. The vessel component is a highly simplified design, and these simplifications will impact the velocity and temperature distribution of the flows within the vessel.

3.7.5 No Radiation Model

The temperatures in this scenario are high enough to generate thermal radiation transfer processes between the hot gas and the system walls or from hot to cooler gas. No thermal radiation model is used for the analysis in this report. Radiation would be expected to reduce the temperature of the gas flowing through the hot leg on its way to the steam generator and could reduce the overall natural circulation flow rates due to the slight reduction in the temperature extremes.

3.7.6 Quasi-Steady Assumption

Fixed boundary conditions are applied to obtain a steady-state solution that represents a snapshot in time of the expected transient behavior. The fixed boundary conditions are obtained from a single point in time selected from SR5 transient predictions. The quasi-steady assumption is valid if the pace of the transient heatup is relatively slow. This assumption is most questionable during the period of rapid core oxidation when the heatup rate is very rapid.

Another issue not addressed by the quasi-steady assumption is the relief valve cycling expected during the transient scenario. All boundary conditions selected from time periods where the relief valves are closed. The experimental evidence at 1/7th scale suggests that the flow pattern reestablishes itself very quickly after the relief valve closes, and the duration of the relief valve opening is short compared to the time during which the relief valve is closed. Sensitivity studies completed using the CFD model also indicated a rapid return to the natural circulation flow pattern after a simulated relief valve opening.

4 BASE-CASE PREDICTIONS AND APPLICATION TO SYSTEMS-CODE MODELS

A best-estimate prediction is made of the natural circulation flows and temperatures based on a simplified set of boundary conditions obtained from a SR5 prediction of severe accident behavior in a four-loop Westinghouse pressurized-water reactor (PWR). This prediction is referred to as the base case because it forms the base from which further sensitivity studies are performed. The CFD predictions provide a detailed look at the flows and temperatures within the upper plenum, hot leg, pressurizer surge line, and the primary side of the steam generator. Specific features of these predictions are integrated out of the predictions and used to refine the models and correlations that are applied in the SR5 model. The predicted results and the applications to a system code model are outlined below.

4.1 Summary of Boundary Conditions

A set of boundary conditions that are representative of the conditions during a severe accident are derived directly from a snapshot of a SR5 transient prediction. The values are obtained shortly after the core uncovers and rapid oxidation begins. Specific values are outlined below.

A boundary condition in the vessel is established to help create the desired severe accident conditions in the vessel upper plenum and to provide the source of mass that leaves the model through the surge line if present. This boundary condition at the base of the vessel creates a small net flow of steam and hydrogen rising up through the core. The flow inlet boundary condition is established on the inner ring of the base of the vessel and is modeled as a velocity inlet. This surface has an area of 1.13065 m^2 and a normal velocity of 0.041284 m/s is established through the surface. Turbulence properties are estimated by specifying a turbulence intensity of 8 percent and a hydraulic diameter of 0.1 m. The temperature of this flow is set to 1,250 K and the hydrogen mass fraction is fixed at 0.0055. These conditions result in a net upward mass flow of 1.3 kg/s which ultimately leaves the model through the pressurizer surge line.

The only outlet in the model is at the end of the pressurizer surge line. The surge line connects horizontally to the hot leg in the base-case model. The vertically mounted surge line is blocked off for this case and is used for a sensitivity study on the surge line orientation. The pressurizer itself is not modeled. Flows into the pressurizer surge line exit the model at a pressure outlet boundary condition. Gauge pressure at the outlet is set to 0.0. The net mass flow through this outlet balances the incoming mass flow from the vessel flow inlet.

The vessel region is utilized in this model to provide a source of additional mass flow into the system as described above and to provide predetermined conditions in the upper plenum region that connects to the hot leg. A fixed temperature condition of 1,250 K is applied to selected regions above and below the upper plenum in the vessel model as a way of controlling the nominal temperatures of the steam-hydrogen mixture in the upper plenum region. No heat transfer or temperature adjustments are made in the upper plenum region. All walls are adiabatic.

The tube bundle is separated into three regions for the purposes of applying boundary conditions. The lowest 0.5398 m of the tube bundle (just above the plenum) is modeled as the tube sheet region. A fixed inner wall temperature of 880 K is used to simulate the temperature of the solid tube sheet wall material over the inlet plenum. The similar region above the outlet plenum is modeled with a fixed wall temperature of 803 K. The tube sheet structure on the outlet side of the steam generator is cooler than the tube sheet material on the inlet side. All of the tubes above the tube sheet are modeled with a convective boundary condition on the outer wall (i.e., secondary side boiler region). An external temperature of 813 K and a heat-transfer coefficient of 86 W/m^2-K is applied to the external surface of the tubes. Stainless steel is used for the wall material and the thickness is set to 0.00127m (0.050 inch). Heat transfer is further augmented using an augmented thermal conductivity. These conditions are one part of an integrated tube bundle modeling approach that is outlined in Appendix A. The tube-bundle model is set up to match the pressure drop and heat-loss characteristics of a prototypical tube bundle as predicted in SR5 modeling of these specific severe accident conditions.

4.2 Solution Overview

The CFD predictions indicate a vigorous natural circulation flow between the upper plenum of the vessel and the steam generator. Figure 8 shows a contour plot of temperature on a vertical plane parallel to the center of the hot leg. Vectors are drawn on the figure to indicate the basic flow directions, and a simplified description of the flow path is provided here. Hot flow from the vessel upper plenum enters the hot leg and flows through the upper hot leg on its way to the steam generator inlet plenum. A portion of this flow is pulled into the pressurizer surge line and exits the model through the outlet at end of the surge line pipe. The hot flow entering the inlet plenum is highly buoyant and accelerates upwards while entraining some of the surrounding cooler gas. A significant flow, approximately twice the flow that entered the hot leg from the vessel upper plenum, enters about 40 percent of the tubes in the bundle and flows to the outlet plenum. This flow is referred to as the hot-tube flow. The hot-tube flow cools significantly before reaching the outlet plenum and then returns to the inlet plenum through the remaining (approximately 60 percent) tubes. This returning flow is referred to as the cold-tube flow. The cold-tube flow enters the inlet plenum and mixes with the hotter flow. This mixing plays a significant role in reducing the temperature of the hot flow that enters the tube bundle. A portion of the cold flow returns to the vessel upper plenum through the lower half of the hot leg completing the natural circulation loop. Some of this cooler flow mixes with the hotter flow in the hot leg as it enters the pressurizer surge line. This mixing helps to lower the temperature of the surge line.

An unsteady interface exists between the hot and cold flows in the hot leg, with vortices shedding along the surface. The majority of the vortex action initiates at the junction to the surge line where hot and cold flows enter the surge line from opposing directions creating a swirling effect. The hot flow along the upper surface accelerates on its way down the pipe, and the interface between the hot and cold flows slopes upwards from the vessel to the steam generator. Some entrainment of the colder flow occurs into the forward-flowing hot stream. The entrainment is greatest as the flow turns upwards into the steam generator inlet plenum. This entrainment increases the mass flow in the forward direction and reduces the average temperature of the hot flow.

A variation in the temperature of the flows entering the tube sheet is quantified. The hottest tube temperature and location varies with time because the buoyant plume in the inlet plenum is unsteady. Generally, the hottest region is directly over the hot leg nozzle exit.

Figure 9 illustrates five slices of the hot leg (labeled 1 through 5) that are used to obtain integrated hot leg data from the predicted results. One key location is slice 1 at the junction of the vessel and hot leg. This surface is used to determine the mass flow and temperature of the flow exiting the vessel. Figure 10 illustrates an example of the predicted results for the temperature and mass flow from the vessel to the hot leg for each of the 140 data sets recorded for the base case. The temperature value oscillates within a band from 1,230 K to 1,233 K. Mass flow oscillations are on the order of +/- 0.5 kg/s. Average values are illustrated on the graph.

4.3 Hot Leg Flow Rates and a Discharge Coefficient

A fundamental issue in the system-code modeling of this type of severe accident scenario is the prediction of the natural circulation flow rates in the hot leg and steam generator tube bundle. Because the system code models do not explicitly model all of the phenomena that govern the natural circulation flow rates, the mass flow in the hot leg and the steam generator tube bundle must be adjusted to match experimental or other alternate model predictions. A model suitable for use in system codes for the determination of the hot leg flow rates is discussed below.

Previous modeling of the severe accident natural circulation flow rates completed for the U.S. Nuclear Regulatory Commission (NRC) using SR5 relied on a process where flow coefficients in the model were adjusted until predetermined conditions were met. These included target values for the ratio of the tube-bundle flows to the hot leg flow (the recirculation ratio) and a specific fraction of core power (around 30 percent) transferred to the steam generators. These conditions were determined experimentally at 1/7th scale. The Advisory Committee on Reactor Safeguards (reference 7) raised concerns about the scalability of the experimental results with respect to the determination of the power fraction transferred to the steam generators. Power fraction is not an ideal scaling parameter in this case. In addition, the experiments did not address core power entering the pressurizer through the surge line. To address these concerns, a new physically based approach for the determination of the hot leg flow rates is suggested that is easier to establish and more consistent in its application.

The proposed approach establishes the hot leg flow rates in a system code model using a discharge coefficient.[10] This method is adapted from a paper by Leach and Thompson[11] that uses scaling arguments to establish that the flow rate is proportional to a densimetric Froude number. The relationship for the volumetric flow rate from Leach and Thompson is provided below.

$$q = C_d \, (g \, D^5 \, \Delta\rho/\rho)^{1/2} \qquad \text{Equation 1}$$

Equation 1 relates the natural circulation flow rate (q - m^3/s) to a constant times the square root of the product of gravity (m/s^2), pipe diameter (m) to the fifth power, and a normalized density difference. The constant term is a dimensionless discharge coefficient, C_d, that is obtained experimentally. This term accounts for geometric features that impact the flow rates and is specific to the geometry of interest. The density term is the difference between the cold and hot densities divided by the average density. This relation provides a means to compute the volumetric flow rate through the hot leg as a function of the geometry and the average densities in the upper plenum and the steam generator inlet plenum. Leach and Thompson demonstrated experimentally that the discharge coefficient remained constant over a fairly wide

19

range of conditions for their given geometry. Their experimental discharge coefficient of 0.09 applied to a straight pipe with an L/D ratio of eight connecting two large tanks.

This approach is applied to the reactor system code models during time periods when the pressurizer power-operated relief valve (PORV) is closed. This is the timeframe where the countercurrent natural circulation flows are established. Flows into the pressurizer surge line continue during this time because the mass in the pressurizer is increasing while the pressurizer PORV is closed. A full understanding of the predicted hot leg flows and the system-code modeling approach is needed before the discharge coefficient can be applied consistently to the system-code model. Differences exist between the CFD predictions and the system-code approach. For instance, the CFD predictions indicate entrainment of cold flow into the hot flow that effectively increases the flow rate along the pipe. This is not possible in the one-dimensional piping arrangement of a system code model.

Table 3 records the average base-case predictions at the five hot leg locations illustrated on Figure 9. At each location, mass-averaged flow and temperature are determined for both directions. Flow from the vessel to the steam generator occupies the upper half of the hot leg, and the return flows are at the bottom. Table 3 data represent the average of 140 data sets recorded at 2-second intervals. Locations 2 and 3 are positioned on either side of the surge line connection, and the mass flow drop between these two faces is used to determine the flow entering the surge line. The average surge line mass flow and temperature is 1.29 kg/s at 1,077K.

The flow rates in the table illustrate the entrainment of the return flows into the hot upper flows. Entrainment between locations 1 and 2 averages 0.67 kg/s. The entrainment between locations 3 and 4 is similar. The most localized entrainment in the hot leg piping occurs in the elbow region where the flow accelerates upwards towards the inlet plenum (between locations 4 and 5). An average of 0.66 kg/s is entrained in this relatively short region. The flow entering the surge line is evident from the mass flow differences between locations 2 and 3. It is observed that 55 percent of the flow into the surge line comes from the hot upper flows and the remaining 45 percent comes from the return flows in the lower hot leg.

Table 3. Hot Leg Flow Rates for Base Case

Hot upper flows from vessel to inlet plenum					
Location	1	2	3	4	5
Mass flow (kg/s)	4.46	5.14	4.42	4.95	5.60
Temperature (K)	1,231.6	1,203.2	1,213.1	1,152.4	1,105.5
Cooler return flows from inlet plenum to vessel					
Location	1	2	3	4	5
Mass flow (kg/s)	3.19	3.85	4.43	4.98	5.67
Temperature (K)	952.5	949.2	954.7	920.5	903.0

The predicted entrainment of cold flow into the hot forward flow is not typically accounted for in the system-code approach. Figure 11 illustrates the hot leg and inlet plenum nodalization from a SR5 model used to predict the flow pattern illustrated in Figure 1. The hot leg is split into two 1D pipes. Pipe 100 carries the hot flow from the vessel upper plenum (volume 582) to the steam generator inlet plenum (volumes 105 and 106). Pipe 101 carries the cooler return flow from the inlet plenum (volumes 106 and 107) to the vessel (volume 581). The inlet plenum volumes (105, 106, and 107) are described in more detail in the discussion of the mixing fraction below. The pressurizer surge line, component 153, is assumed to be a side-mounted design and draws mass from both the upper and lower hot leg pipes.

For application of Equation 1 to the system code model for a hot leg from Figure 11, the following assumptions are made. The hot density, ρ_h, is obtained from cell 582 that feeds the upper hot leg. The flow conditions entering the hot leg from cell 582 are assumed to represent the average conditions in the upper plenum. Similarly, the properties of the flow from the inlet plenum to the lower hot leg are used to determine the cold density (ρ_c). A mass-weighted cold density is obtained from the flows entering the lower hot leg from inlet plenum volumes 106 and 107. The density difference and average density are computed from these values. The full hot leg diameter is used in the correlation (not the smaller diameter of the individual upper and lower pipes). The volumetric flow rate, q, is determined by dividing the mass flow rate by the average density. A net mass flow from the vessel to the steam generator is used, and this value is obtained at the junction between volumes 100-04 and 100-05. These values are used in Equation 1 along with a predetermined value of the discharge coefficient to test for consistency with Equation 1. Flow-loss coefficients can be adjusted to ensure Equation 1 is satisfied. The value of C_d is predetermined from applicable experiments or from CFD predictions as outlined in this report.

The CFD predictions in Table 4 are utilized as a numerical experiment to determine the discharge coefficient in Equation 1. The discharge coefficient is determined to be 0.126 from the base-case predictions, and the value is found to be relatively steady. The value found is representative of the geometry in Figure 7. For significantly different geometries, a different discharge coefficient would need to be determined. To determine the discharge coefficient for the specific geometry, the hot and cold flow densities at each end of the pipe as well as a representative flow rate is needed. The hot density, ρ_h, is obtained from a mass-averaged integral of the forward flow entering the hot leg from the vessel (position 1 in Figure 9). This provides the density of the flow from the reactor vessel to the hot leg and is consistent with the assumptions used above for the system-code model. Similarly, the mass-averaged density (ρ_c) of the flow from the steam generator inlet plenum into the lower hot leg is obtained as an integral of the return flow at position 5 (Figure 9). A volumetric flow rate is obtained from a mass flow rate and the average density. The mass flow rate of interest represents the net mass flow from the vessel to the steam generator. This is obtained by subtracting ½ of the surge line mass flow from the mass flow entering the hot leg from the vessel at position 1 (Figure 9). The subtraction of ½ of the surge line flow is based on the approximation that half of the surge line flow comes from the upper hot leg flow. This assumption was determined to be reasonable over a wide range of conditions. The resulting mass-flow determination is consistent with the method described above for the system-code approach. It also is noted that the system-code model should be set up to ensure that the mass flow into the surge line comes equally from the upper and lower hot leg pipes. This is outlined below in the surge line discussion.

The preceding discussion lays out an approach to modeling the hot leg flow rates in system codes such as SR5 or MELCOR using a physically based Froude number correlation. The CFD predictions are used to determine a discharge coefficient, C_d, which in turn can be used to establish the hot leg mass flow in system-code models. The average discharge coefficient predicted from base-case CFD predictions, averaged over a total of 140 data sets, is 0.126. To

21

give some indication of the variation in this parameter, the 140 data sets are broken down into 7 consecutive groups of 20 data sets each (40-second intervals). The discharge coefficients computed for each of the 7 data groups (0.125, 0.128, 0.129, 0.122, 0.128, 0.119, and 0.130) range from 0.119 to 0.130, which represents less than a 6-percent variation from the mean. The standard deviation computed from these values is 0.004.

4.4 Mixing Fraction, Recirculation Ratio, and Hot Tube Fraction

The 1-dimensional nature of the pipe flows in the system-code approach requires some type of mixing model to account for the mixing and entrainment of the flows that has been observed experimentally and predicted with multidimensional CFD codes. Historically, the mixing has been assumed to occur primarily in the steam generator inlet plenum. Figure 12 illustrates the inlet plenum mixing model[12] that has been applied for many years. The recirculation ratio is defined as the ratio of the tube bundle mass flow to the hot leg mass flow. A mixing fraction is defined that refers to the fraction of flow from the hot leg that mixes completely in the inlet plenum. These terms are typically determined experimentally. Another feature that needs to be predetermined, although not directly part of the mixing model, is the hot-tube fraction. Hot-tube fraction defines the fraction of tubes that carry hot flow from the inlet to outlet plenum.

Figure 12 outlines the mass and energy exchanges considered in the mixing model. Mass flow from the upper hot leg, m, is split by the mixing fraction. A portion represented by f enters the central mixing volume (T_m). The remainder of the flow, 1-f, bypasses the mixing volume and proceeds to the steam generator tube bundle. The same mixing fraction is assumed for the flow returning from the tube bundle through the cold tubes. A relation for the mixing fraction, f, and the mixed inlet plenum temperature, T_m, is obtained by applying mass and energy conservation for a steady-state, steady-flow process to the control volumes in Figure 12. The resulting relations are listed below in Equations 2 and 3.

$$f = 1 - r \, (T_{ht} - T_m) \, / \, (T_h - T_m)$$ Equation 2

$$T_m = (T_h + r \, T_{ct}) \, / \, (r + 1)$$ Equation 3

$$r = m_t \, / \, m \quad \text{(recirculation ratio)}$$

The approach assumes that experimental methods will provide T_h, m, T_{ct}, T_{ht}, and m_t. These values are then used to determine values for the mixing fraction and recirculation ratio. The values of T_h and m are typically measured near the steam generator end of the hot leg, and values for T_{ct}, T_{ht}, and m_t, are measured at the tube sheet face (tube entrance or exit). Flow coefficients are adjusted in the system code models to establish the flow rates between volumes to ensure that the mixing fraction and recirculation ratio are consistent with the predetermined values. CFD predictions (reference 6) also have been used like a numerical experiment to predict T_h, T_{ct}, T_{ht}, and r. CFD predictions provide valuable insights that can extend the available experimental results to new geometries or conditions.

In the present analysis, T_{ct}, T_{ht}, and m_t are determined by integrating the CFD predictions for the tube-bundle flows at the tube-sheet face. These values are easy to define and straightforward to calculate from the CFD results. The values for T_h, and m, however, require a little more consideration. The data in Table 4 indicate that the mass-flow changes along the hot leg pipe and entrainment in the hot leg reduces the temperature of this flow. This mixing occurs prior to the inlet plenum and is not considered by the mixing model approach described earlier. A problem with the approach described above is that that mixing fraction and recirculation ratio are somewhat dependent upon the location where T_h and m are determined due to the hot leg mixing.

A new approach to the mixing model is suggested that accounts for the mixing in the hot leg and the inlet plenum. Values for T_h and m are determined at the vessel end of the hot leg (location 1 in Figure 9). These values represent the temperature and mass flow leaving the vessel upper plenum and can be clearly defined. This approach expands the mixing region to include the hot leg as illustrated in Figure 13. This new mixing region includes the pressurizer surge line connection, if present. Figure 14 illustrates a crude representation of the mixing model volumes corresponding to a system-code model. The assumption of a steady-state, steady-flow process is made and the conservation laws are applied as before. The resulting equations for f and T_m are listed in Equations 4 and 5. The equations are similar to the previous mixing model with the exception of the recirculation ratio which is modified to account for flows into the surge line (m_s) if present. The main difference from the previous approach is in the definition of where the hot temperature, T_h, is defined.

$$f = 1 - r\, (T_{ht} - T_m) / (T_h - T_m)$$ Equation 4

$$T_m = (T_h + r\, T_{ct}) / (r + 1)$$ Equation 5

$$r = m_t / (m - m_s/2) \quad \text{(recirculation ratio)}$$

The benefit of this mixing model is the clear definition of terms (i.e., what is T_h and m) and the consistency with the predicted hot leg and inlet plenum flow behavior. Accounting for the hot leg mixing and entrainment results in an increase in both the mixing fraction and recirculation ratio compared to the earlier approach.

Averaged values from the 140 data sets of the base-case prediction (T_{ct} = 841.2 K, T_{ht} = 961.4 K, m_t = 9.12 kg/s, T_h = 1,231.6 K, and m = 3.81) are used to determine an average mixing fraction and recirculation ratio. The average and standard deviation for these values, along with the average prediction for the hot-tube fraction, are listed below. Figure 15 shows the individual mixing fraction and recirculation ratio data for each of the 140 individual data sets.

Mixing fraction	f = 0.95 +/- 0.16
Recirculation ratio	r = 2.4 +/- 0.3
Hot tube fraction	44 % +/- 3% (3% of unplugged tubes)

The implementation of these mixing parameters in a system code model is an iterative process. Multiple flow coefficients must be adjusted in the system code models to ensure consistency with the CFD predictions. The monitoring points are described with respect to the typical noding diagram outlined in Figure 11. The tube-entrance temperature, T_{ht}, is obtained as a mass-averaged temperature coming from inlet plenum volumes 105 and 106. The tube-exit temperature, T_{ct}, is simply the temperature of the cold-tube volume 111-26. The recirculation ratio is defined as the tube-bundle mass flow divided by the net hot leg flow. To obtain an average value, the hot and cold flows are averaged for both the hot leg and tube bundle. An

23

average tube-bundle mass flow is obtained from the tube-bundle entrance flow (volumes 105 and 106) and exit flow (from volume 111-26). The average net hot leg mass flow is obtained from the hot leg exit flows (to volumes 105 and 106) and the hot leg entrance flows (from volumes 106 and 107). The value for T_h (the hot leg hot-flow temperature) is obtained from cell 100-05 at the end of the hot leg. This value is taken at this end of the hot leg to allow for the heat losses to the pipe wall that were not included in the steady-state CFD predictions. This value does not impact mixing and entrainment in the hot leg since the 1-dimensional upper hot leg is isolated from the lower hot leg. Hot-tube fraction is established in the system codes through the assignment of flow areas to the hot- and cold-steam generator tubes. This parameter is fixed in the model.

4.5 Tube Variations and the Hottest Tube

A typical system-code model for the steam generator includes only 1 hot tube, which represents all of the tubes carrying hot flow from the inlet to outlet plenum. Figure 11 shows a typical noding diagram for a SR5 model, and the first cell of the hot tube (component 110) is shown as control volume 110-01. Flow enters the hot tube from the inlet plenum at a single mass averaged temperature (T_h). Experimental results and the CFD predictions, however, show a significant temperature variation across the hot-flow tubes. The extent of this temperature variation, and specifically the hottest tube temperature, is an important consideration for the integrity of the steam-generator tubes.

Figure 16 shows predicted temperature contours for one data set from the base-case results on a horizontal cross section at the elevation of the tube sheet entrance. A boundary is drawn to illustrate the boundary between the hot and cold (return) flow tubes. The return tubes are at a temperature near the secondary side temperature. The boundary separating the upflow (hot) and downflow (cold) tubes changes occasionally at the edges in a random manner. The hot upward flow appears to lock in on certain tubes for a period of time and then occasionally goes through a transition that results in a new overall pattern. These changes mainly affect the outer edges of the boundary and are most significant in the regions furthest from the hot leg nozzle. The central hot tube region, where the hottest average tubes are located, remains in an upflow condition during the entire simulation. A boundary is drawn around the 21 hottest tubes (average) on Figure 16. This region is symmetric around the single average hottest tube which is also highlighted. During the transient simulation, the position of the specific hottest tube changes by several tubes in each direction (referring to the enlarged tube pitch of the CFD model) so the absolute hottest temperature is not always applied to the same tube.

The tube entrance temperatures are normalized to make it practical to superimpose the temperature variations on the system code predictions. A histogram of the fraction of tubes that are subjected to specific temperature ranges is used to determine the highest temperatures that enter the tube bundle and the number of tubes subjected to this temperature. The tube inlet temperature, T_{ht}, is normalized using Equation 6.

$$T_n = (T_{ht} - T_{ct}) / (T_h - T_{ct})$$ Equation 6

The values of T_h and T_{ct} (see Figure 13) represent the hot leg inlet temperature and the cold-tube return flow temperature, respectively. These are the hottest and coldest temperatures associated with the hot leg and inlet plenum mixing. The normalized temperature, T_n, is bound between zero and one. A normalized temperature of one indicates no mixing between the hot leg and the tube-sheet entrance. A normalized temperature of zero represents flow into the tube bundle at the temperature of the flow returning to the inlet plenum from the cold tubes.

The normalized temperature range is divided into 20 ranges, and the fraction of tubes within each range is obtained for each of the 140 data sets in the base-case prediction. Figure 17 provides the determination of an average value and standard deviation for the fraction of tubes for each range as well as the plotted results. These data provide the average fraction of tubes within a given temperature range. Considering the data in the temperature range from 0.5 to 0.55, the results indicate that an average of 0.6 percent of the tubes (0.6 percent of the unplugged tubes) fall into this range. This represents 2 tubes in the CFD model or 18 tubes in a full-scale steam generator that has 3,388 tubes. The location of the two tubes in this range varies with time. For this reason, the normalized temperatures presented in this manner provide a conservative upper bound for the normalized hottest tube entrance temperatures. The data in the range from 0.65 to 0.7 represent less than 1/9th of a single tube in the CFD model (less than 1 tube in a full-scale generator) and are considered insignificant. The highest range with a tube fraction that represents more than 1 full scale tube is the 0.6 to 0.65 range where an average of 0.16 percent of the tube bundle is predicted to be at any given time. This approach is similar to the method used in NUREG-1788 (reference 6) to predict the hottest tube ranges and the results are consistent for the hottest tube. These results, however, do not account for the fact that the hottest temperatures are not always applied to the same tube.

Figure 17 is conservative since the hottest tube location changes. No single tube is predicted to have a sustained normalized temperature range of 0.6 to 0.65. Figure 18 shows a plot of the normalized tube entrance temperatures for the 10 hottest average tubes. The 140 individual points are connected by lines with symbols for every 10th point. The data show that the hottest average tubes do not stay at the hottest temperatures for extended periods of time. These same tubes also see below-average temperatures. This plot makes it clear that the upper bound of normalized temperatures obtained from Figure 17 represents a conservative upper bound on the hottest tube temperature.

A more realistic approach is adopted to find the hottest tube temperatures by considering the tubes individually. Figure 19 shows a histogram of the normalized tube temperatures for all tubes considered on an individual basis. A mass weighted average normalized temperature is determined for each tube in the model, and these values then are collected into the 20 temperature ranges defined earlier. The spread of the data is reduced compared to Figure 17 because the peaks of the fluctuations are averaged out. This plot indicates that the hottest tubes have a long time average normalized entrance temperature that falls within the 0.4 to 0.45 range. The single hottest tube averaged over the 140 data sets is found to have a normalized temperature of 0.43 +/- 0.1 (0.1 is the standard deviation). Breaking these data down into 7 groups containing 20 data sets each (7- 40 second intervals) results in average normalized temperature predictions for the hottest tube of 0.51, 0.38, 0.44, 0.45, 0.44, 0.39, and 0.43. These predictions may slightly under predict the prototypical hottest temperature due to the enlarged size of the tubes (i.e., the CFD model uses one tube to represent nine prototypical tubes). Combining tubes in this way will have a tendency to average out the peaks. The impact of combining 9 tubes is expected to be small but is noted since it is not a conservative assumption.

The purpose of this section is to provide information for estimating the hottest tube temperatures in the bundle because it is these hottest tubes that experience the greatest thermal challenge during a severe accident scenario. Two approaches are presented. The normalized maximum tube inlet temperature in the 0.60 to 0.65 range, as suggested by Figure 17, represents a conservative upper bound. A more realistic normalized temperature range that considers each tube individually (Figure 19) suggests a hottest average normalized tube temperature in the 0.4 to 0.45 range. The average hottest tube is found to have a normalized average temperature of 0.43 with a standard deviation of 0.1. This value comes from a series of seven 40 second intervals where the highest average value over one of the intervals is 0.51.

Since the hottest tube prediction fluctuates around, the CFD model was further executed with a focus on the 21 hottest tubes and data were obtained at a higher frequency. A total of 240 seconds of additional run time was simulated and data were output every 0.1 seconds for a total of 2400 data points. The resulting average normalized hottest tube temperature was 0.43 over all of these data which is consistent with the prior prediction. Once again, the data were broken down into 40 second intervals and the maximum normalized tube temperature for one of these intervals was found to be 0.5. This too is consistent with the initial prediction. The thermal time constant of the tube wall is estimated to be 30 seconds so results for these 40 second time intervals may be significant. In light of the results outlined above, along with a consideration of the uncertainty in the predictions, it is recommended that a value of 0.5 be used to represent the normalized hottest tube temperature in a system-code analysis for a Westinghouse plant.

To estimate the temperature of the hottest tubes from system-code predictions, Equation 6 is solved for T_{ht} with the value of T_n set to the maximum predicted value (i.e. 0.5). The value of T_h should come from the steam generator end of the hot leg (cell 100-05 in Figure 11). This cell provides the hot leg hot flow temperature after the heat loss to the hot leg wall is taken into account. The cold tube temperature, T_{ct}, is obtained from the last cell in the cold steam generator tube component (cell 111-26 in Figure 11). In a recent set of SR5 predictions completed by NRC, a side calculation is completed for a single tube that represents the hottest tube (normalized temperature of 0.5). This calculation is performed outside of the overall system model as a side calculation. The system code model predicts the average conditions of the hot tubes and the side calculation provided the estimate of the hottest tube behavior. The inlet temperature for this hottest tube calculation is set to T_{ht} as determined above. The other flow and boundary conditions are established to match the hot average tube conditions from the system code model (component 110 in Figure 11).

4.6 Surge Line Flows and Mixing

The surge line for the base-case model is attached to the side of the hot leg. The CFD predictions indicate that the average flow into the surge line comes nearly equally from the upper and lower hot leg flows. Figure 20 shows the percentage of mass flow into the surge line from the upper (hot) and lower (cold) hot leg flows for the base-case prediction. The hot forward flow in the hot leg contributes an average of 55 percent of the surge line mass flow and the remaining 45 percent comes from the cooler return flow. Given the variations observed in this data and the knowledge gained from additional sensitivity studies, a 50:50 mixture assumption is considered appropriate for system-code modeling applications.

This ratio must be forced on system codes with hot leg noding similar to that illustrated by Figure 11. It is noted that the upper and lower hot legs are not connected in the SR5 model, and the pressure in cells 100-03 and 101-03 are not representative. Cell 101-03 is essentially far downstream of cell 100-03 in the piping network used to model the natural circulation flows. Simply connecting these two cells to the surge line (cell 153-07) results in an unbalance of flow rates. The majority of flow will come from the upper hot leg pipe (100-03) that is at a higher pressure. Control logic is needed in the system code models to enforce a 50:50 mixture into the surge line consistent with the 3-dimensional model predictions.

4.7 Stochastic Nature and Average Values

The pattern of hot tubes in the tube bundle appears random at the edges, and the initial conditions are suspected of having an impact on the initial pattern as it is established. To test this, two additional base-case predictions are made using two different sets of initial conditions. Qualitatively, the predictions are all very similar. The overall flow patterns and general locations of the hottest tubes are unchanged. The predictions did indicate some variation in the hot tube pattern at the boundaries and a small change in the mixing parameters of interest. Table 4 lists the average predictions for the base case and the two additional predictions. The final column lists the average of the three sets of simulations. The repeated cases were limited to 68 and 40 data sets, which do add to their uncertainty.

Table 4. Base-Case Repeatability and Average Values

		Base Case	Repeat 1	Repeat 2	Average
Number of data sets (2 s intervals)		140	68	40	-
C_d	Discharge coefficient	0.126	0.121	0.122	**0.12**
F	Mixing fraction	0.95	0.98	0.96	**0.96**
R	Recirculation ratio	2.4	2.4	2.3	**2.4**
Hot tube fraction		44 %	41 %	39 %	**41 %**
T_n	Normalized T (hottest tube)	0.43	0.39	0.45	**0.43**
Surge line flow split (hot:cold)		55:45	51:49	46:54	**51:49**

The predictions are not fully repeatable, partially because of the stochastic nature of the hot plume and upflow tube pattern. In addition, there is acknowledgement that the limited number of data points used to determine the averages does impact the results. Sensitivity studies using SR5 demonstrate that the variations predicted for these mixing parameters do not significantly impact the system code modeling predictions. The most significant variable from a tube integrity perspective is the normalized hottest tube temperature. This value is predicted to have longer term oscillations and recording data for 40 or 80 seconds is not sufficient to obtain a repeatable average value. In order to obtain valid predictions for the normalized hottest tube temperature, each simulation completed was executed for an additional extended period of time with the 21 hottest tube temperatures recorded at 0.1 second intervals. These extended results were used to determine the normalized hottest tube temperatures.

Another long term variation that is observed is the fraction of tubes that carry hot flow. In one long term simulation example, it was observed that over extended periods of the simulation, the tube-flow patterns occasionally changed. For significant periods of time the upflow pattern is unchanged and then, at some point, the flow pattern adjusts at the edges and another tube-flow pattern is established. In this particular example, the fraction of hot tubes changed significantly from 32 to 43 percent. In the system-code modeling, this would impact the temperature predicted for the average hot tube. The more important consideration, however, is how this change impacts the hottest tube region where the tubes are most likely to fail. In this example, the normalized mass-averaged temperature entering the hottest tube did not change when the tube-flow pattern adjusted from 32 to 43 percent. This suggests that variations in the tube upflow pattern do not impact the hottest tubes at the core of the upflow region. These hottest tubes are the principal concern. The use of the long term average for the fraction of hot tubes is considered acceptable for system code modeling.

5 SENSITIVITY STUDIES

A series of sensitivity studies are completed to gain a better understanding of the severe accident natural circulation flows in the reactor coolant system and the parameters that can impact them. The base-case results outlined above use best-estimate boundary conditions to determine a set of mixing parameters and other settings for use in a system-code modeling approach. The following sensitivity studies consider variations in the input parameters and other modeling assumptions in order to demonstrate how these conditions impact the important mixing parameters of interest in the base-case model. The range of parameter variation is not tied directly to an expected range of variation in any specific system code analysis and it is used as a demonstration only. It was found that 40 to 80 seconds of transient simulation data are required to obtain consistent results for the mixing parameters of interest but the base case predictions demonstrate that more data may be needed to obtain a consistent average value for the hottest tube prediction. Typical sensitivities use 40 to 80 seconds of full data sets to determine average mixing parameters. Each of these cases is then executed for an extended period of time with data collection focused on the 21 hottest tubes to ensure a reliable hottest tube prediction.

5.1 Disruption and Reestablishment of Natural Circulation Flows

The CFD predictions are focused on the severe accident natural circulation flow pattern illustrated by Figure 1. The CFD predictions are obtained using steady-state boundary conditions and assume a quasi-steady behavior in the overall system. System-code predictions and available test data show that this flow pattern is disrupted when a pressurizer relief valve is opened for a short time to relieve system pressure. During this transient valve opening, the flow into the surge line is significantly increased and the countercurrent flow pattern is disrupted. A prediction is made to consider the time it takes to reestablish the natural circulation flow pattern after the relief valve is closed. The base-case prediction is continued with an increased flow (30 kg/s) into the surge line. The assumption is made that 20 kg/s comes from the vessel side of the hot leg and 10 kg/s comes from the steam generator. These boundary conditions set up a flow pattern where flow enters the hot leg from both ends and exits through the surge line connection. Once this new pattern is completely established, the boundary conditions are returned to the base-case values and the return to countercurrent flow is monitored.

Figure 21 shows the incoming hot leg flow rate and temperature as a function of time starting at the point where the boundary conditions are returned to the base-case values. The base-case values also are indicated on the plot. It takes roughly 10 seconds for the incoming hot leg mass flow to recover to the steady-state values. The temperatures take longer to settle out. The steam generator tube-bundle mass flows are not as quick to respond. Tube-bundle rates ramp up to over 70 percent of the original value within 20 seconds. Beyond 20 seconds, the tube-bundle flows begin to asymptotically approach the steady-state values.

These qualitative results indicate a rapid reestablishment of the natural circulation flow pattern. The tube-bundle flows take about twice as long as the hot leg flows to reestablish. These results are considered qualitative, however, due to the model limitations for predicting transient behavior. The core model is not set up to predict the transient core and upper plenum behavior during and after the pressurizer relief valve cycles. In addition, the steam generator tube-bundle model does not include the thermal inertia of the steam generator tubes. Having these hot

29

tubes in place is expected to speed up the reestablishment of the tube-bundle flows. The predictions obtained with the current model are considered only a rough indication of the speed at which the flow rates can be reestablished after a pressurizer relief valve is cycled.

5.2 Increased Hydrogen Mass Fraction

A sensitivity study is completed to consider the impact of the hydrogen mass fraction. The hydrogen is assumed to be completely mixed with the steam in the system. Adding hydrogen reduces the density of the overall mixture. The base-case model assumed a hydrogen mass fraction of 0.0055. This mass fraction is doubled to 0.011 and then doubled again to 0.22 for a second sensitivity. Doubling the hydrogen mass fraction reduces the mixture density by 4.3 percent at a temperature of 1,000K. Doubling the hydrogen again (to four times the base case) to 0.022 reduces the mixture density by 11.9 percent from the base case (at 1000K).

Table 5 lists the calculated parameters of interest for the increased hydrogen case along with the average of the base-case predictions. The results are generally consistent with the base-case prediction but do indicate slightly more mixing which is evidenced by the reduced hottest tube temperature. The drop in the normalized hottest tube temperature and the slight increase in the recirculation ratio are the same for both sensitivity studies. The reduced temperature is consistent with the slightly higher mixing and entrainment suggested by the increased recirculation ratio. These predictions indicate that hydrogen variations of this magnitude have a modest impact on the hottest tube temperatures.

Table5. Sensitivity Study on Hydrogen Mass Fraction

		H2 x 2	H2 x 4	Base Case - Average
H2	Mass fraction	0.011	0.022	**0.0055**
C_d	Discharge coefficient	0.12	0.12	**0.12**
f	Mixing fraction	1.0	0.97	**0.96**
r	Recirculation ratio	2.5	2.5	**2.4**
	Hot tube fraction	43 %	48 %	**41 %**
T_n	Normalized T (hottest tube)	0.38	0.38	**0.43**
	Surge line flow split (hot:cold)	49:51	50:50	**51:49**

5.3 Surge Line Mass Flow, No Surge Line

A second set of sensitivity studies are completed to look at the impact of the mass flow into the surge line. The surge line mass flow is increased by 30 percent, decreased by 30 percent, and reduced to zero (by eliminating the surge line). These changes did not have a significant impact on the discharge coefficient even though the hot leg mass flow rate is affected. The predicted mass flow entering the hot leg from the vessel for the +30 percent, -30 percent, and no surge line cases is 4.3 kg/s, 3.8 kg/s, and 3.6 kg/s, respectively. This indicates that the mass flow rate leaving the vessel increases with increasing surge line mass flow as expected but that a single value for the discharge coefficient can be used to model this behavior over the given range. The mixing fraction shows no significant variations and there is only a slight impact on the recirculation ratio which changes by less than 10 percent. A peculiar result is the slightly lower normalized temperature for the hottest tube for all cases. There is no clear trend with the magnitude of the surge line flow and all results are within 7 percent of the base case. The low hot-tube fraction for the +30-percent mass flow case is considered an anomaly. This type of tube-flow pattern has been seen in other cases only to later switch to a larger fraction at a later point in the simulation. This particular case appears to have locked onto a small number of hot tubes and remained steady for the period of time when data were collected. As noted earlier, the reduced fraction of hot tubes does not appear to impact the temperatures of the hottest tubes. For all cases, the surge line flow split remained close to 50:50.

Table 6. Surge Line Mass Flow Sensitivity

		+ 30%	- 30%	No surge line	Base Case-Average
m_s	Surge line mass flow (kg/s)	1.69	0.91	0	**1.3**
C_d	Discharge coefficient	0.12	0.12	0.12	**0.12**
f	Mixing fraction	1.0	0.95	0.98	**0.96**
r	Recirculation ratio	2.2	2.5	2.5	**2.4**
	Hot tube fraction	31 %	43 %	44 %	**41 %**
T_n	Normalized T (hottest tube)	0.41	0.40	0.40	**0.43**
	Surge line flow split (hot:cold)	49:51	47:53	N/A	**51:49**

The results for no surge line indicate that the mixing parameters and discharge coefficient are applicable with reasonable accuracy for loops with and without the pressurizer.

5.4 Top-Mounted Pressurizer Surge Line

In some plants, the pressurizer surge line is connected to the top of the hot leg. The base-case model uses a side-mounted surge line connection. Flows into the surge line are expected to come mainly from the upper hot leg flows in this case, and this is expected to lower the temperature of the flow entering the tubes. To quantify this, a prediction is made using the base-case model with the surge line connection rotated up to the top of the pipe. Figure 2 shows the CFD model with both the side- and top-mounted surge lines visible. Only one surge line pipe (or none) is used at for any given simulation.

Table 7 shows the predictions for the top-mounted surge line. No significant differences occur in the discharge coefficient, recirculation ratio, or the hot tube fraction when the surge line is mounted from the top. Two significant differences are predicted. First, the surge line flow split ratio shows about ¾ of the flow entering the surge line comes from the upper hot leg (hot) flow. It was assumed that all of the flow into the surge line would come from the upper hot leg, but the predicted unstable nature of the hot leg flows results in some flow from the lower portion of the cold leg (cold return flows) entering the surge line. The second difference observed follows from the prediction that the top-mounted surge line removes more upper hot leg flow compared to the side-mounted surge line of the base case. The result is a lower normalized hottest tube temperature. Table 7 shows the average for the hottest tube and the average of the hottest range overall are both reduced by almost 0.1. These results indicate that the risk of severe accident-induced failure of a top-mounted surge is higher than for a similar side-mounted design. As a corollary, the results indicate that a top-mounted surge line will decrease the risk of a severe accident-induced steam generator tube failure on the affected loop compared to a loop with a side-mounted or no surge line.

The mixing fraction is not computed for the top-mounted surge line case because the surge line split ratio is not 50:50 as assumed in the mixing model formulation. Assuming a 50:50 surge line split results in a mixing fraction slightly over 1.0. For the purposes of system code modeling, a mixing fraction of 1.0 is recommended for a loop with a top-mounted surge line. The difference between a mixing fraction of 0.96 (base case) and 1.0 is negligible in system code predictions.

Table 7. Sensitivity Study – Top-Mounted Surge Line

		Top-Mounted Surge Line	Base Case - Average
C_d	Discharge coefficient	0.11	**0.12**
f	Mixing fraction	1.0 (recommended)	**0.96**
r	Recirculation ratio	2.6	**2.4**
	Hot tube fraction	44 %	**41 %**
T_n	Normalized T (hottest tube)	0.34	**0.43**
	Surge line flow split (hot:cold)	76:24	**51:49**
T_n	Normalized T (hottest range**)	0.50 – 0.55	**0.6-0.65**

* see Figure 17 and discussion under section 4.4

5.5 Steam Generator Secondary Side Temperature

The relatively cool temperature at the secondary side of the steam generator provides the ultimate heat sink for this model. A change in this temperature changes the rate of heat transfer through the tube walls and, with the associated impact on the fluid temperatures, changes the overall buoyancy driving forces in the natural circulation loop. Table 8 shows the results to two sensitivity studies where the secondary side temperature was changed by +/- 100 degrees.

Table 8. Steam Generator Secondary Side Temperature Sensitivity

		+ 100K	- 100K	Base Case-Average
T_{ss}	Secondary side T (K)	913	713	**813**
C_d	Discharge coefficient	0.127	0.119	**0.12**
f	Mixing fraction	0.98	1.0**	**0.96**
r	Recirculation ratio	1.75	2.81	**2.4**
	Hot tube fraction	26 %	49 %	**41 %**
T_n	Normalized T (hottest tube)	0.47	0.36	**0.43**
	Surge line flow split (hot:cold)	47:53	54:46	**51:49**

** Actually computed to be slightly over 1.0. Value rounded down to 1.0 for system code model.

The change in the temperature has a significant impact on the mass flow in the tubes. When the temperature is raised by 100 degrees, the buoyancy driving force in the tube bundle is reduced and the mass flow in the tube bundle changed from 8.6 kg/s (base case) to just under 4.7 kg/s. Alternatively, reducing the secondary side temperature increased the buoyancy driving force and the tube-bundle mass flows increased to 12.9 kg/s. The mass flows in the hot leg showed similar changes with the highest mass flow predicted for the lowered temperature case.

The increased mass flows in the tube bundle associated with the reduced temperature case results in a higher recirculation ratio. A higher proportion of cold flow is entrained into the hot flow that enters the tube bundle. This reduces the temperature of the flow entering the tubes as indicated by the decreased normalized temperature value and the mixing fraction that is slightly higher than 1.0. The increased flow rate in the tube bundle results in a hot tube fraction that approaches 50 percent. Ignoring the thermal impact on the pressure drop in the tube bundle, a 50-percent hot tube fraction is a natural result of the flow attempting to find the path of least resistance for high flow rates.

The predictions associated with the increased secondary side temperature indicate lower mass flows in both the hot leg and tube bundle. A lower proportion of cold flow is entrained into the hot flow as indicated by the reduced recirculation ratio. Although the mixing fraction remains essentially unchanged compared to the base case, the hottest tube prediction indicates less mixing. The reduced hot tube bundle flows occupy only 26 percent of the tube bundle. These results indicate that the steam generator secondary side temperatures have a significant impact on the hot tube fraction, the recirculation ratio, and the hottest tube temperatures. The prediction of the discharge coefficient shows less than a 6 percent variation.

5.6 Reactor Vessel Upper Plenum Temperature

This sensitivity study considers the impact of the average upper plenum temperature that is defined here as the mass-averaged temperature of the flow from the upper plenum to the hot leg. A change in this temperature changes the total temperature difference (overall buoyancy force) that drives the natural circulation flows. Table 9 shows the results of two sensitivity studies where the upper plenum temperature was changed by approximately +/- 100 degrees.

Table 9. Reactor Vessel Upper Plenum Temperature Sensitivity

		+ 100K	- 100K	Base Case-Average
T_h	Vessel upper plenum T (K)	1,327	1,138	1,232
C_d	Discharge coefficient	0.115	0.125	0.12
f	Mixing fraction	1.0	0.94	0.96
r	Recirculation ratio	2.3	2.4	2.4
	Hot tube fraction	33 %	47 %	41 %
T_n	Normalized T (hottest tube)	0.37	0.43	0.43
	Surge line flow split (hot:cold)	54:46	44:56	51:49

The change in the upper plenum temperature has an impact on the predicted modeling parameters as outlined in Table 9. The mixing fraction, recirculation ratio, and discharge coefficient showed only minor variations from the base case. The hot tube fraction variations are not consistent with those observed for the secondary side temperature sensitivity outlined in section 5.5 for a given overall temperature difference between the vessel upper plenum and the steam generator secondary side. This inconsistency is attributed to the somewhat random behavior of the tube flow pattern as noted earlier. The normalized hottest tube temperature for the elevated temperature case, but this value is within the normal variations observed for the base case predictions. Overall, no significant sensitivity is predicted to the upper plenum temperature.

5.7 Pressure Drop in Steam Generator Tube Bundle

A sensitivity study is completed to consider the impact of the pressure drop correlations used to augment the tube-bundle pressure drop. The coefficients (see Appendix A) that are used to establish the tube-bundle pressure drop are changed by +/- 30 percent. A summary of the results is contained in Table 10.

Table 10. Tube Bundle Pressure Drop Sensitivity

Tube Bundle Loss Coefficients		+ 30%	- 30%	Base Case-Average
C_d	Discharge coefficient	0.12	0.12	**0.12**
f	Mixing fraction	0.98	0.99	**0.96**
r	Recirculation ratio	2.3	2.5	**2.4**
	Hot tube fraction	40 %	34 %	**41 %**
T_n	Normalized T (hottest tube)	0.38	0.37	**0.43**
Surge line flow split (hot : cold)		47:52	53:47	**51:49**

The change in the pressure-drop modeling has only a minor impact on most of the mixing parameters of interest. The discharge coefficient is unaffected. As expected, the higher pressure drop reduces the tube bundle flows and the recirculation ratio is slightly affected. A similar change in the recirculation ratio is predicted for the lower pressure drop case. The variations in the mixing fraction are not considered significant. A puzzling result is the drop in the normalized temperature of the hottest tube for both cases. No explanation is found for both variations resulting in a reduced temperature. The models were re-run for an extended period of time with a focus on the hottest tube temperatures and the results were repeated.

5.8 Leakage From Steam Generator Tubes

A qualitative sensitivity study is completed to look at the effect of leakage from a local region near the hottest tube location. Leakage rates of 1.5, 3.0, 6.0, and 12.0 kg/s are considered. For each case, the tube-entrance surface at the tube-sheet face is modeled as an outlet. It is assumed that 90 percent of the leakage comes from the inlet plenum and 10 percent of the leakage flow comes through the tube from the outlet plenum. This is only an assumption, and no modeling is completed to determine a best-estimate distribution of the leakage flows.

34

A single leaking tube (which represents a 3 x 3 array of tubes in a prototype steam generator) is set up to loose mass at the lower tube sheet face. Figure 3 identifies the tube location. This lower tube sheet surface is selected for the leak location because a surface was already defined in the model at this location. In hindsight, a tube-leakage location should have been created further up in the tube bundle during the model development. The modeling impact of removing the mass as soon as the flow encounters the tube entrance is not considered to be significant in light of the other assumptions already adopted for this tube leakage sensitivity. The tube leakage location is set up to remove mass from both sides of the surface so that some mass is removed from the outlet plenum side of the surface. The system-code mixing parameters and other coefficients are not computed for these leakage sensitivity studies since the mixing model formulations do not account for a leak. Some information can be learned from the hottest tube temperatures around the leak. In addition, a qualitative description of the leakage impact on the natural circulation flows is provided.

For the case with a leakage rate of 1.5 kg/s, the overall flow pattern is very similar to the base case. Surprisingly, no apparent shift occurs in the average normalized temperature for the hottest tube, and the fraction of tubes carrying hot flow is unchanged. Qualitatively, the flow pattern looks similar to the base-case prediction. The rising plume and the hot leg countercurrent flows remain unstable and the leak did not attract the plume directly. Even the flow-split ratio for flow into the pressurizer surge line remains consistent with the base case. Leaks of 1.5 kg/s or less under these conditions do not appear to have a significant impact on the overall flows and mixing in the loop.

The second case added a tube leakage rate of 3.0 kg/s to the base case. In this prediction, the countercurrent flow pattern persists, but the return flow to the vessel is greatly reduced. The hot flow occupies a larger portion of the hot leg. Flows into the pressurizer surge line are approximately 75 percent from the upper hot leg (hot) flow. The plume appears to have some attraction to the leak location, but the instabilities of the flow pattern are more significant and the plume continues to be move around. The hottest average tube increases by about 0.05 on a normalized scale for this case. Leaks of this magnitude have a modest effect on the overall flow and temperature distribution in the loop.

The leakage flow rate is increased to 6.0 kg/s for the third sensitivity study. In this prediction, the countercurrent flow pattern almost goes away. A small return flow on the bottom of the hot leg is occasionally backed up by the forward flow. The average cold return flow in the hot leg was is than 0.15 kg/s. Flows into the surge line are predominantly (90 percent) from the hot forward flow. The hot rising plume in the inlet plenum still demonstrates an unsteady nature. Occasionally, the plume is found to lock onto the leaking tube. Although the flow in the hot leg is almost one directional, a strong natural circulation flow pattern remains in the steam generator and this helps to cool the rising plume. The average temperature for the hottest tube (not counting the leaking tube) is found to be approximately 0.1 higher than the base case on a normalized scale. Intermittent peaks in the temperature reached 0.8 on the normalized scale. Leakage rates of this magnitude begin to break down the countercurrent flow in the hot leg and result in a significant decrease in the mixing.

A final sensitivity study is completed using a leakage rate of 12.0 kg/s. The hot leg flow is completely in the direction of the steam generator under these conditions. No return flows are predicted. Strong flows still exist in the steam generator tube bundle, and the rising plume in the inlet plenum still displays some unsteadiness and does mix to some degree. The plume spends more time attracted to the leaking tube in this case. The average temperature of the flow reaching the hottest tube (next to the leaking tube) is 0.8, and the tube-entrance temperature occasionally reaches 1.0 on the normalized scale. This indicates very little mixing in some regions. Leakage rates of this magnitude result in a direct flow from the vessel to the steam generator. Some of the hot flow is drawn up into the tube bundle with very little mixing.

6 HOT LEG AND SURGE LINE CONVECTIVE HEAT TRANSFER

Convective heat transfer to the hot leg and surge line is another issue that is modeled with the system codes. Some qualitative CFD work was completed to study this issue, and some observations and recommendations for system code modeling are outlined below. The CFD predictions of hot leg and steam generator natural circulation flows, which are the subject of this report, assumed adiabatic walls for the hot leg and surge line pipes. Attempts to add heat transfer to the hot leg were unsatisfactory. Setting up accurate initial conditions to represent a snapshot of the transient heating of the pipe was difficult. The predictions completed are only considered qualitatively. One thing was clear from the results. The predicted heat transfer rates to the hot leg and surge line wall were higher than those reported by the system codes. This led to further study of the issues and the recommendations outlined below.

The limited CFD modeling of hot leg heat transfer indicated mixed convection heat transfer on the upper hot leg walls. A secondary flow pattern in the hot leg was clearly predicted. As the steam mixture cooled near the hot leg wall, it fell downward and therefore reduced the thickness of the developing thermal boundary layer. Predicted heat-transfer rates were significantly higher than those computed by typical fully developed heat-transfer correlations. A search for data found no heat transfer measurements at the Reynolds numbers and temperature differences of interest, and the CFD results were considered only qualitatively. The principal concern is that the system codes, with their fully developed heat transfer correlations, will under-predict the heat transfer to the hot leg pipe. It is suggested that entrance effects could be considered to justify enhancing the convective heat transfer coefficients. The issue of mixed convection correlations will enhance the heat transfer

The SR5 system code model of the hot leg pipe utilizes the Dittus-Boelter[13] correlation that applies to the heating of pipe walls by convection. This correlation gives the Nusselt number (Nu) that is used to determine the convective heat transfer coefficient along the horizontal section of the hot leg and surge line. The correlation and related parameters are provided below.

$Nu_{DB} = 0.023 \, Re^{0.8} \, Pr^{0.3}$ Dittus-Boelter correlation for Nusselt Number

$h_c = Nu_{DB} \, k \, / \, L$ convective heat transfer coefficient

Re = Reynolds Number
Pr = Prandtl Number
k = thermal conductivity of the fluid
L = hydraulic diameter used for pipe flows

The Dittus-Boelter correlation is based upon fully developed pipe-flow data. The hot leg and surge line flows, however, are far from fully developed. A typical hot leg in a nuclear power plant has a length to diameter (l/d) ratio of less than 10. The area of concern in the surge line is at the entry point where the development length is effectively zero. Flow patterns begin to approach fully developed conditions at l/d values greater than 15 and in some cases require l/d values near 50. Measured local Nusselt numbers in the entry region of a circular tube are considered by Kays and Crawford.[14] The data show that the local Nusselt number within the first seven pipe diameters of flow length is about 1.2 to 2.4 times higher than the fully developed value. Table 11 provides the ratio of the local to fully developed Nusselt number for various flow length-to-diameter ratios in the entry length of a tube flow as taken from the plot in Kays and Crawford.

Table 11. Local Nusselt Number* in Entrance Region

l/d	1	2	3	4	5	6	7	8	9	10
Nu_x/Nu	2.38	1.85	1.54	1.39	1.30	1.23	1.19	1.15	1.12	1.10

For system-code modeling of the severe accident flows, entrance effects are recommended to be considered when evaluating the convective heat-transfer coefficient in the hot leg and surge line. The values in Table 11 could be used if no specific data are available.

In a recent study (reference 4) using SR5 to model a four-loop Westinghouse plant, the entrance effects were applied to the computation of the convective heat-transfer coefficient in the hot leg and surge line. During periods of countercurrent flow (see Figure 9), the heat transfer of the upper hot leg flows was enhanced using the entrance effects data. The hydraulic diameter used in the correlation assumed that the flow occupied ½ of the pipe. Increasing the heat transfer to account for the entrance effects increases the heating of the hot leg pipe and reduces slightly the temperature of the steam that reaches the tube bundle. In a similar manner, the heat transfer to the surge line also should be updated to remove the bias of using a fully developed heat-transfer correlation in the entrance region. The hot leg is typically predicted to fail near the end of the nozzle region, which is only 1 or 2 diameters from the entrance where the entrance effects are most significant. Similarly, the surge line failure is expected right at the entrance region where the surge line connects to the hot leg. The author believes that issues such as mixed convection and turbulence in the entrance regions could further enhance the heat transfer beyond what is outlined in Table 11. This topic is left for future study.

---

* Adapted from Kays and Crawford, Figure 13-10, sharp entry

7 SUMMARY OF RESULTS AND RECOMMENDATIONS

The completed analysis uses an improved CFD model to determine mixing parameters and coefficients for tuning a system code model applied to severe accident simulations with three-dimensional (3D) natural circulation flows. The CFD model used in this study encompasses a series of lessons learned from several years of analyses including a benchmark study at 1/7th scale (reference 5) and a follow-on study of full-scale steam generators (reference 6). The updated modeling also addresses Advisory Committee on Reactor Safeguards comments on those earlier studies (reference 7).

The natural circulation flows between the reactor vessel upper plenum and the steam generator are predicted under specific severe accident conditions that are obtained from prior system-code model predictions. A vessel model establishes the conditions in the upper plenum, which feeds the natural circulation flows in the hot leg, pressurizer surge line, and the primary side of a steam generator. A countercurrent flow pattern is established that carries heat from the upper plenum to the steam generator tube bundle. An unsteady buoyant plume is predicted in the inlet plenum as the hot steam and hydrogen mixture rises up and into the tube bundle. Time-averaged mass flows and temperatures are obtained throughout the system, and these predictions are used as a numerical experiment to define flow and mixing parameters for use in tuning a system-code model.

A modified mixing formulation is established to account for the hot leg and inlet plenum mixing as well as the pressurizer surge line flows. This updated formulation is considered to be an improvement over earlier models that focused solely on the inlet plenum mixing. In addition, a discharge coefficient is defined that can be used to predict the hot leg mass flow rates based on the densities in the vessel upper plenum and the steam generator inlet plenum. The predictions provide a means of tuning a system code to obtain the mass flows and temperature distribution in the hot leg, surge line, and steam generator tube bundle. These predictions can be used to extend the existing experimental data into the specific steam generator geometry and severe accident conditions studied.

The recommended system code modeling parameters for a Westinghouse plant (assumed to have a model 51 steam generator) or plant with similar steam generator designs are summarized below.

$f = 0.96$	Mixing fraction
$r = 2.4$	Recirculation ratio
41 %	Hot tube fraction
$C_d = 0.12$	Discharge coefficient
$T_m = 0.5$	Bounding normalized temperature of hottest tube
50 : 50	Hot : Cold flow split ratio into side mounted pressurizer surge line

Sensitivity studies are completed to provide an estimate of the variation in these parameters under a variety of conditions and assumptions. In all cases, the discharge coefficient remained relatively constant with maximum variations of less than 8 percent. This demonstrates the benefits of using this approach to establish the hot leg flows in a system code model. Similarly, the mixing fraction is found to vary by only a few percent over the range of conditions considered. The recirculation ratio is found to be sensitive to the secondary side temperature.

Although not considered in this study, the tube bundle heat-transfer rate was found to impact the recirculation ratio in previous work (reference 6). The temperature and heat-transfer rates in the tube bundle affect the buoyancy driving forces. These parameters are found to have the largest impact on the recirculation ratio. The value suggested above, 2.4, is obtained using conditions pulled directly from a realistic system code prediction of severe accident conditions in a Westinghouse pressurized-water reactor (PWR).

The hot tube fraction is used for sizing the hot and cold steam generator tube sections in a system code model. This parameter is difficult to predict with confidence because some of the tubes at the margin (i.e., tubes at the edge of the hot and cold regions) seem to occasionally change direction and the hot tube fraction can change by 10 percent or more in a given analysis. The predictions were not carried out long enough to obtain a consistent long-term average value. One important finding is that the hottest tube region does not appear to be significantly affected by changes in the overall size and shape of the hot tube region. In other words, the core of the hot tube region is somewhat consistent. Changes to the tube flow patterns occur at the edges of the hot tube region where the temperatures are more moderate. The base-case prediction had a longtime average hot tube fraction of 0.41. This value is in the middle of the range of all of the predictions. When the tube bundle flow is significantly increased, the hot tube fraction apparently tends to approach 0.5. At the lowest tube bundle flow rates predicted, the hot tube fraction is found to be as low as 0.26.

The normalized temperature of the hottest tube is a significant parameter because it refers to the portion of the tube bundle where the thermal loading is most severe. This parameter has been utilized in recent NRC studies (reference 4) for the purposes of determining whether a tube will fail prior to the hot leg or some other RCS component. In the base-case prediction, the mass-averaged normalized temperature entering the hottest tube is found to be 0.43. This value fluctuates around. The data sets were broken down into 40-second intervals, and the study found that the normalized temperature reached 0.5 over some of these intervals. For this reason, a value of 0.5 is recommended as a bounding value for system code models. The sensitivity of this parameter to changes in the modeling parameters was significant. Average values ranging from 0.36 to 0.47 were obtained. The most significant variation came from changes in the secondary side temperatures. A separate sensitivity study that moved the surge line to the top of the pipe also showed a significant impact on the hottest tube temperature. The top-mounted surge line removes some of the hottest flow, and the average normalized temperature of the hottest tube drops to 0.34.

The flow (hot:cold) split ratio into the surge line pipe is predicted for simulations that included a pressurizer surge line. This variable remained generally within 5 percent of a 50:50 split ratio over the range of sensitivity studies, and a 50:50 split ratio is recommended for system code models with a side-mounted surge line. The temporal variations in this parameter were very large and indicated significant turbulent fluctuations at the surge line to hot leg connection. The 50:50 value represents a long-term average value. The one sensitivity that did significantly impact this result involved moving the surge line to the top of the hot leg. In this case, approximately 75 percent of the flow into the surge line came from the hot flow in the upper pipe section. The top-mounted surge line therefore is subjected to a larger thermal challenge than a side-mounted surge line. This could be important in cases where the surge line is predicted to fail prior to the hot leg.

The series of predictions completed with a range of tube leakages from the primary to secondary side help to quantify the significance of tube leakage on the overall natural circulation flows. A leakage rate of 1.5 kg/s resulted in no significant variation. The countercurrent natural circulation between the vessel upper plenum and the steam generator is maintained for leakage rates up to 6 kg/s but, as the leakage rates increase, the average temperature of the flow entering the tube bundle increases. For a leakage rate of 12 kg/s, the countercurrent flow pattern is essentially broken and the steam temperatures entering the tube bundle begin to approach the hot leg (hot flow) temperatures.

Some prior qualitative CFD results highlighted the fact that some system-code models will underpredict the convective heat-transfer rates to critical regions of the hot leg and surge line. In the regions where the thermal boundary layer is still developing, the fully developed heat-transfer correlations used in system codes underpredict the heat-transfer rates. To account for this underprediction of convective heat transfer, a set of factors are provided that can be used to adjust the fully developed heat-transfer correlation to account for the local entrance region effect. These factors, or other data if more appropriate, should be applied in the determination of the hot leg and surge line convective heat-transfer rates. In addition to the thermal entrance effects, it is expected that much of the upper hot leg also will experience mixed convection that would further increase the convective heat transfer to the hot leg. This topic is suggested for future research if a more detailed analysis of the hot leg becomes necessary.

These CFD predictions build on prior studies (NUREG-1788) and incorporate improvements and lessons learned in the analysis. The results outlined here are considered to be best-estimate predictions for use in system-code model tuning for severe accident natural circulation flows in a Westinghouse PWR. As shown in the prior work (NUREG-1788), the steam generator design impacts the flows and mixing during this type of scenario. The results outlined here are not universal to all reactor types. The lessons learned, however, should provide insights into any future analyses in this area.

41

8 REFERENCES

1. Thadani, Ashok and Collins, letter to Samuel, memorandum outlining the steam generator action plan, U.S. Nuclear Regulatory Commission, May 11, 2000.

2. U.S. Nuclear Regulatory Commission, "Severe Accident Natural Circulation Studies at the INEL," NUREG/CR-6285, INEL-94/0016, February 1995, Agencywide Document Access and Management System (ADAMS) Accession No. ML071220028.

3. U.S. Nuclear Regulatory Commission, "Risk Assessment of Severe Accident-Induced Steam Generator Tube Rupture," NUREG-1570, March 1998, ADAMS Accession No. ML003769765.

4. U.S. Nuclear Regulatory Commission, "SCDAP/RELAP5 Thermal-Hydraulic Evaluations of the Potential for Containment Bypass During Extended Station Blackout Severe Accident Sequences in a Westinghouse Four-Loop PWR," NUREG/CR-6995, July 2009.

5. U.S. Nuclear Regulatory Commission, "CFD Analysis of 1/7th Scale Steam Generator Inlet Plenum Mixing During a PWR Severe Accident," NUREG-1781, October 2003, ADAMS Accession No. ML033140399.

6. U.S. Nuclear Regulatory Commission, "CFD Analysis of Full-Scale Steam Generator Inlet Plenum Mixing During a PWR Severe Accident," NUREG-1788, May 2004, ADAMS Accession No. ML041820239.

7. Advisory Committee on Reactor Safeguards, letter to Travers, William D., U.S. Nuclear Regulatory Commission, May 21, 2004, ADAMS Accession No. ML041420237.

8. J.D. Anderson, *Computational Fluid Dynamics: The Basics with Applications*, McGraw-Hill Series in Mechanical Engineering, McGraw Hill, 1995.

9. Nuclear Energy Agency, "Best Practice Guidelines for the Use of CFD in Nuclear Reactor Safety Applications," NEA/CSNI/R(2007)5, Paris, France.

10. C.F. Boyd, "Prediction of Severe Accident Counter Current Natural Circulation Flows in the Hot Leg of a Pressurized Water Reactor" (U.S. Nuclear Regulatory Commission), *Proceedings of ICONE 14, International Conference on Nuclear Engineering, 17-20 July 2006*, Miami, Florida.

11. S.J. Leach and H. Thompson, *"An Investigation of Some Aspects of Flow Into Gas Cooled Nuclear Reactors Following an Accidental Depressurization,"* J. Br. Nuclear Energy Society, July 14, 1975, No. 3, 243-250.

12. D. B. Ebeling-Koning, et al., "Steam Generator Inlet Plenum Mixing Model for Severe Accident Natural Circulation Conditions," 1990 ASME Winter Meeting, Dallas, Texas.

13. Pitts, D.R., Sissom, L.E., "Shaun's Outline Series, Theory and Problems of Heat Transfer," pg.170, McGraw Hill Inc., 1977.

14.	Kays W.M., Crawford M.E., "Convective Heat and Mass Transfer", 2nd Edition, Figure 13-10, McGraw Hill, 1980.

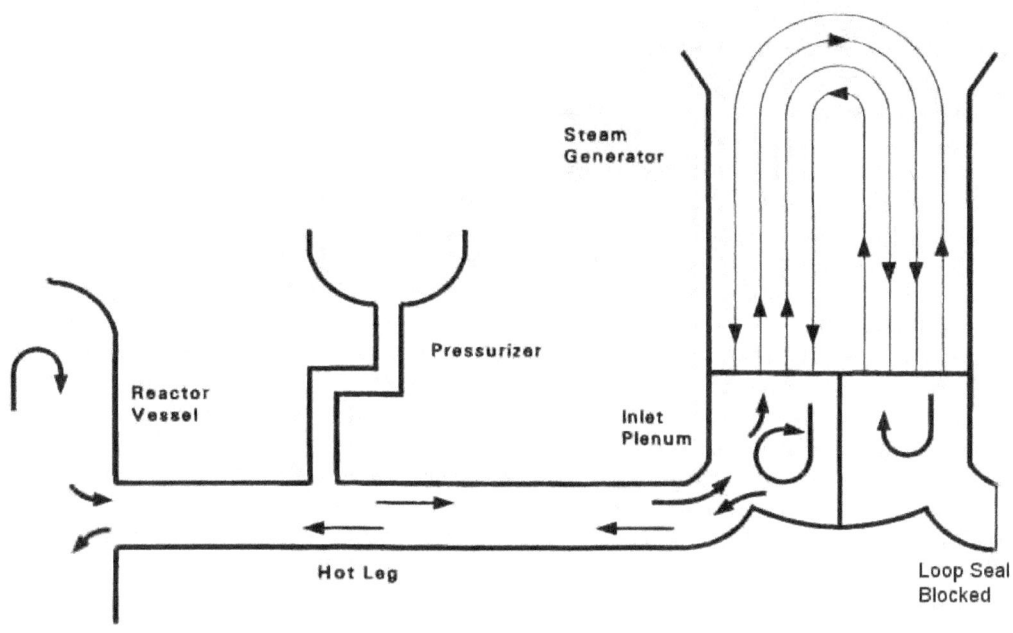

Figure 1. Severe Accident Natural Circulation Flow Pattern of Interest

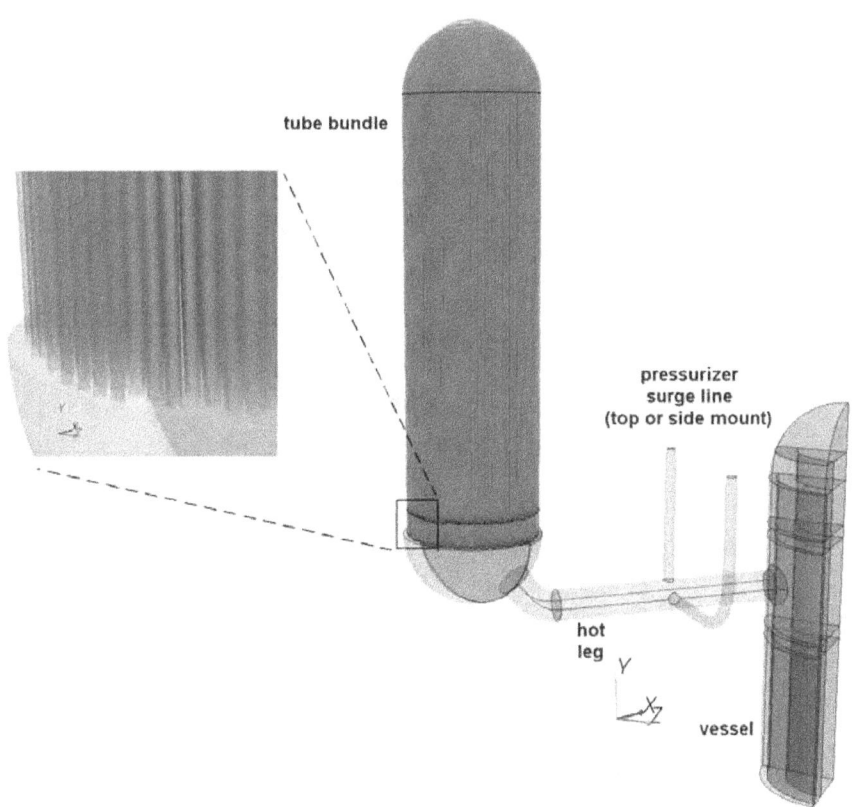

Figure 2. Overview of CFD Model Domain

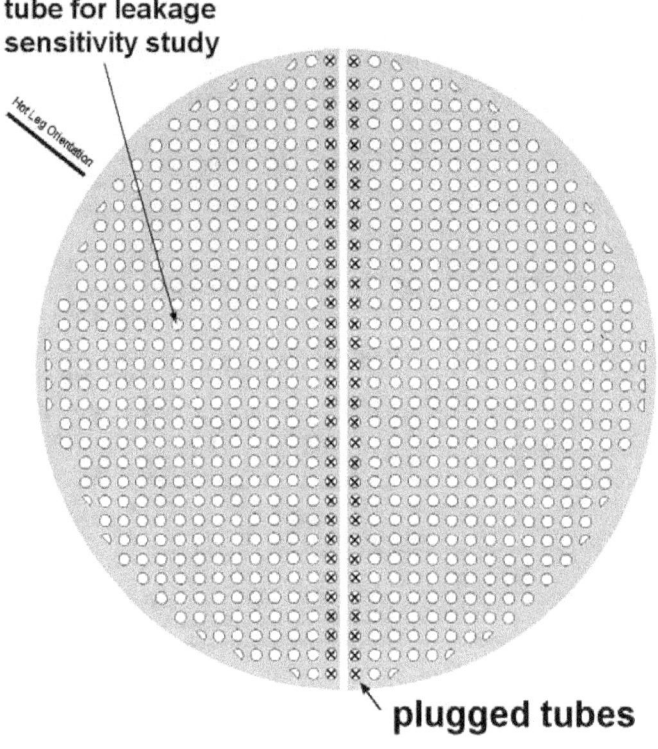

Figure 3. Tube Bundle Cross Section

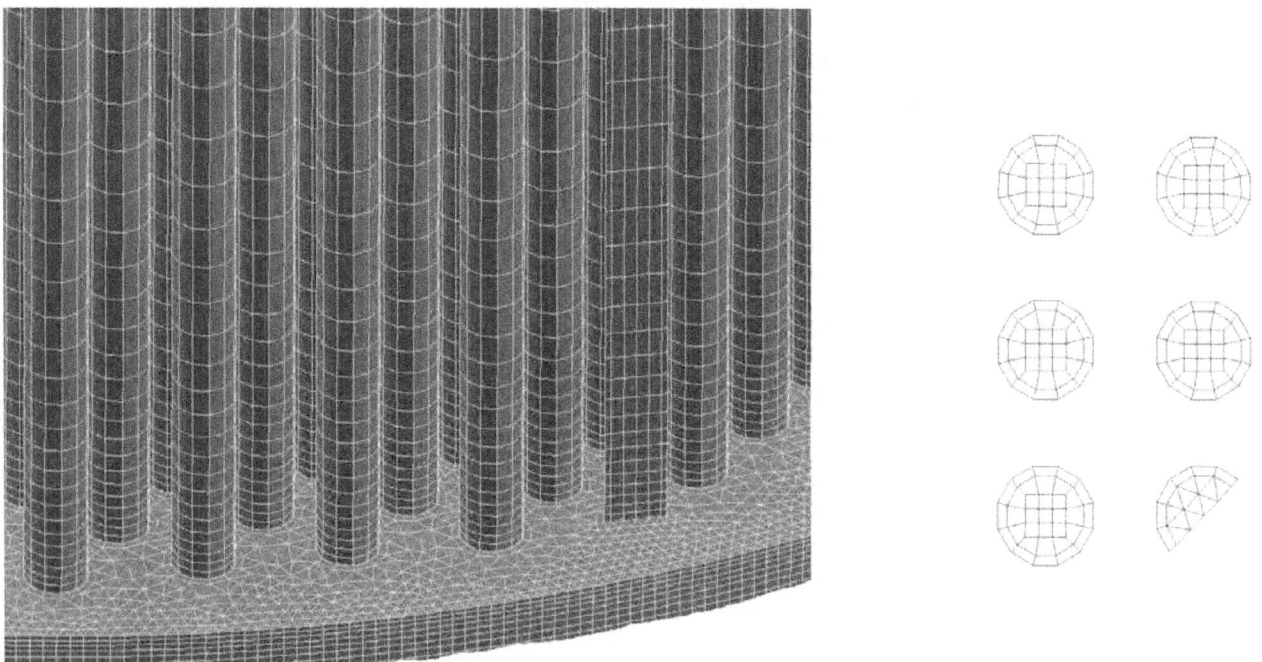

Figure 4. Tube Bundle Mesh Details

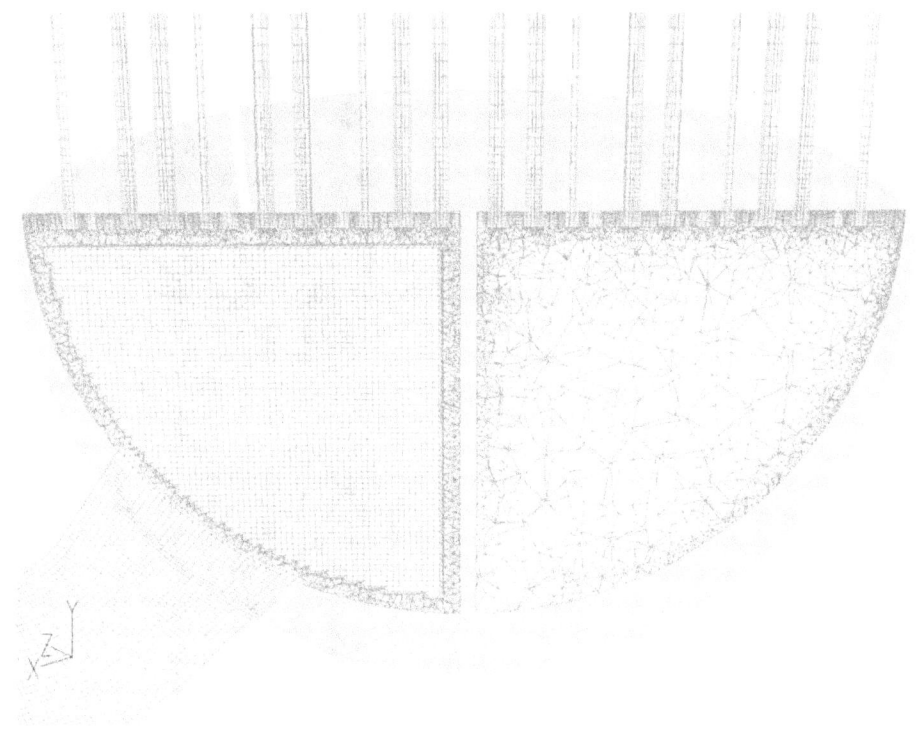

Figure 5. Mesh Cross Section through Steam Generator Plenums

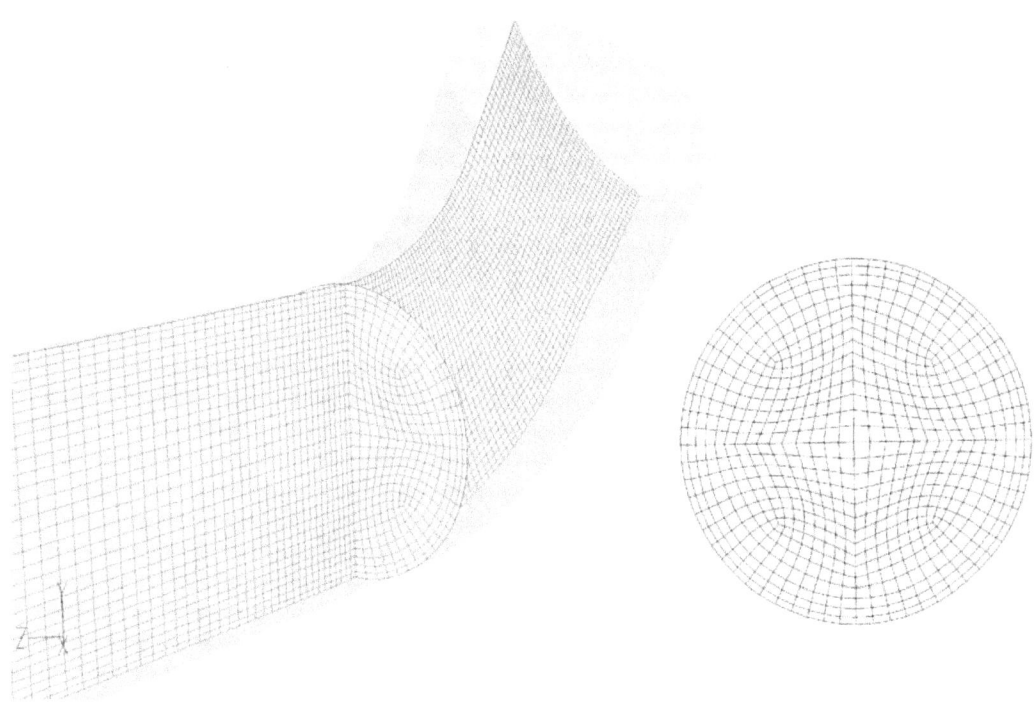

Figure 6. Hot Leg Mesh Details (Centerline and Cross Section)

47

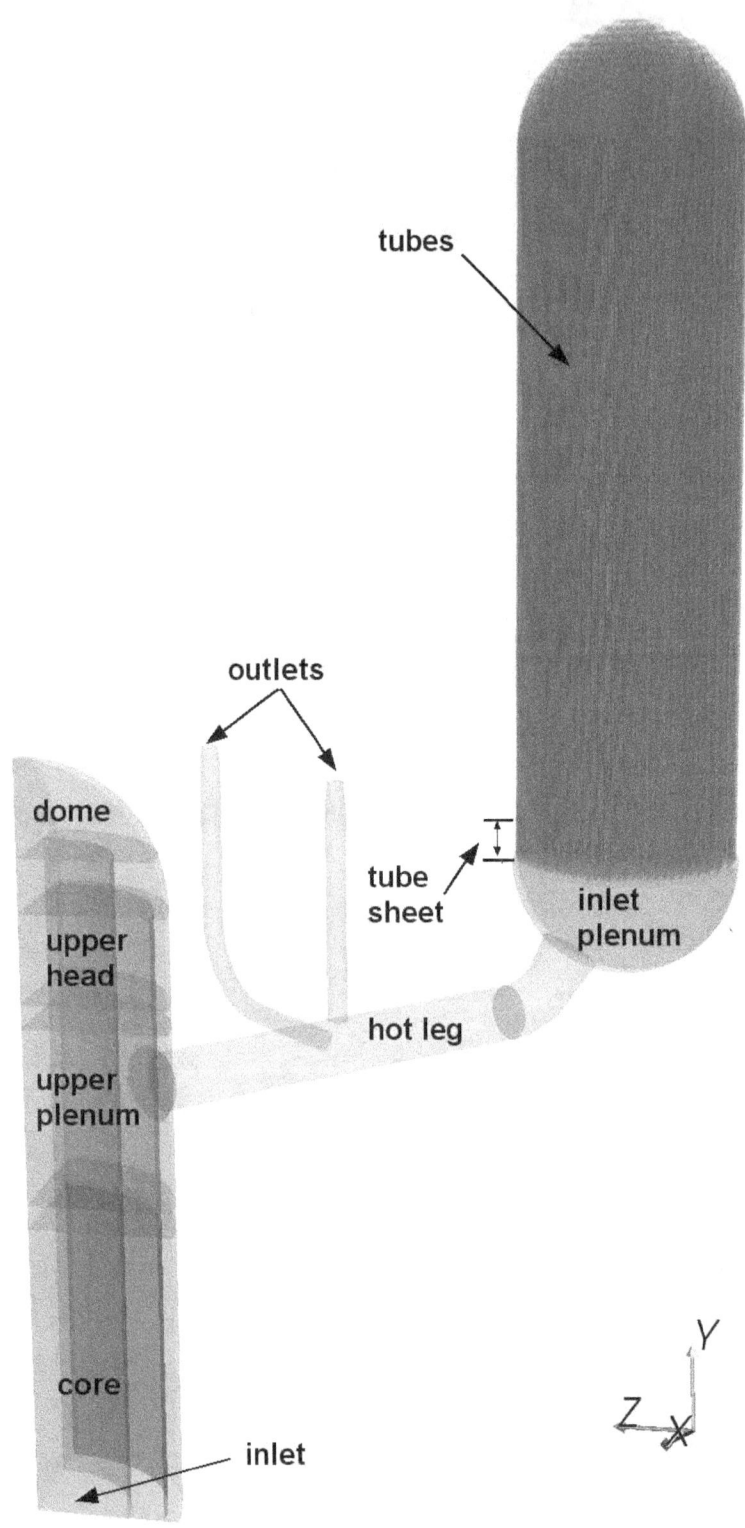

Figure 7. Boundary Condition Regions for CFD Model

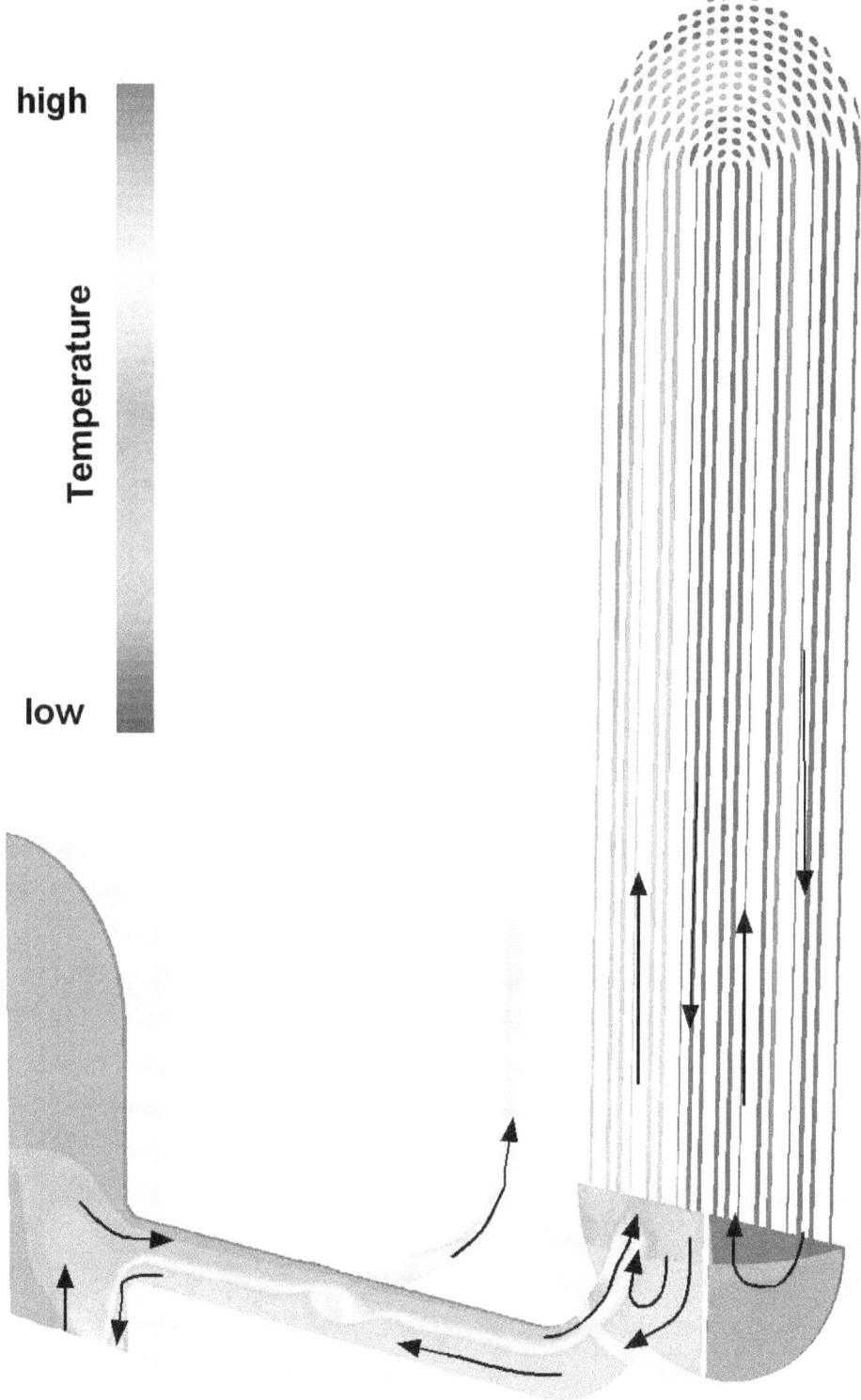

Figure 8. Temperature Contour Patterns and Flow Patterns on Hot Leg Centerline Plane

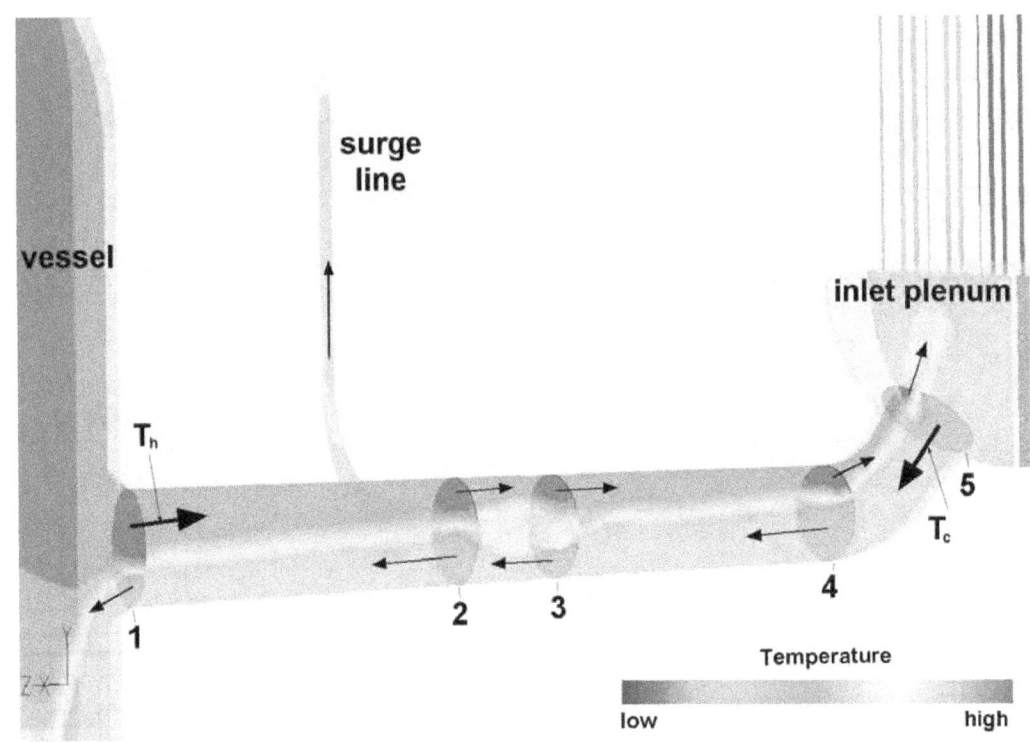

Figure 9. Temperature Contours Indicating Hot Leg Pattern with Analysis Locations

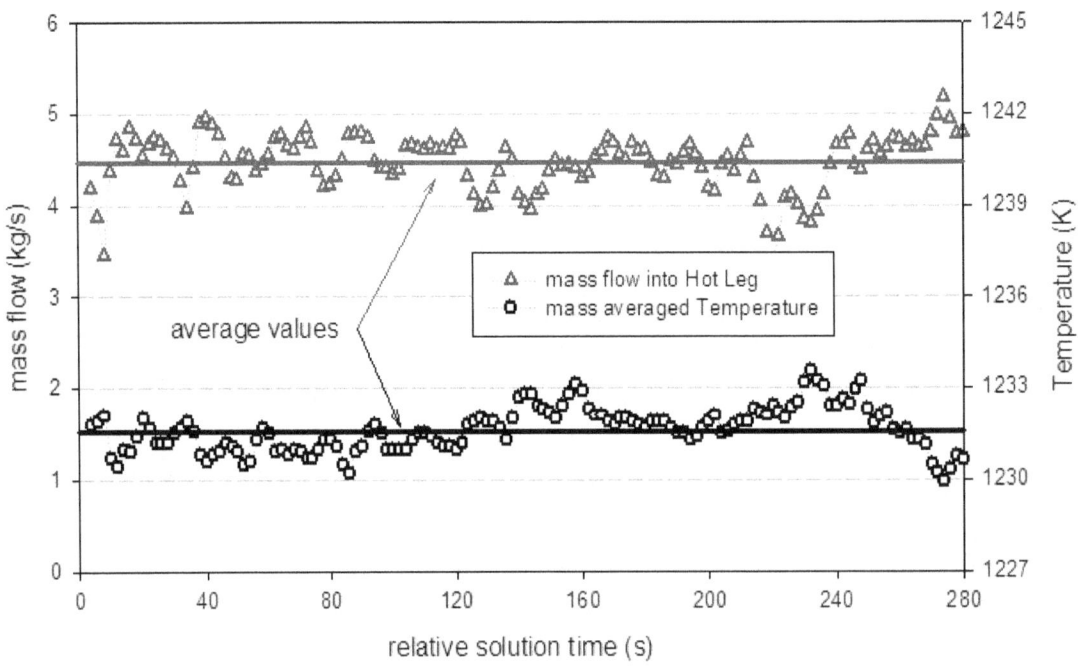

Figure 10. Temperature and Mass Flow Entering Hot Leg from Upper Plenum

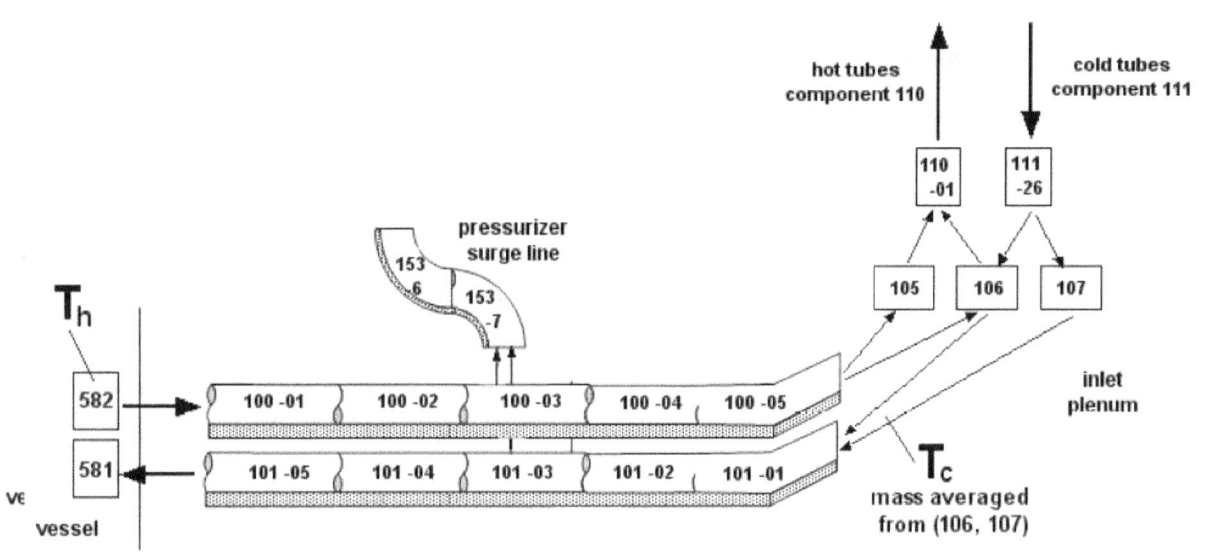

Figure 11. A SCDAP/RELAP5 System Code Nodalization of Split Hot Leg and Inlet Plenum

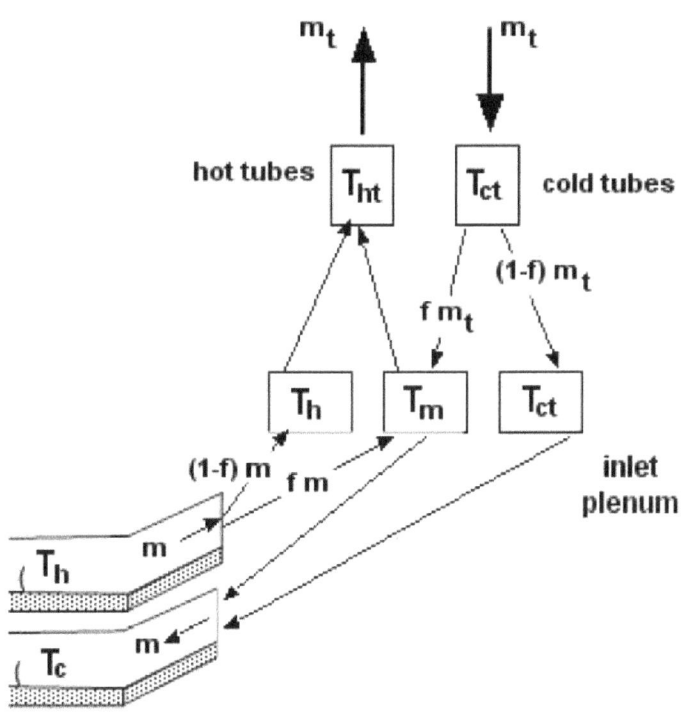

Figure 12. Diagram of Inlet Plenum Mixing Model for System Codes

51

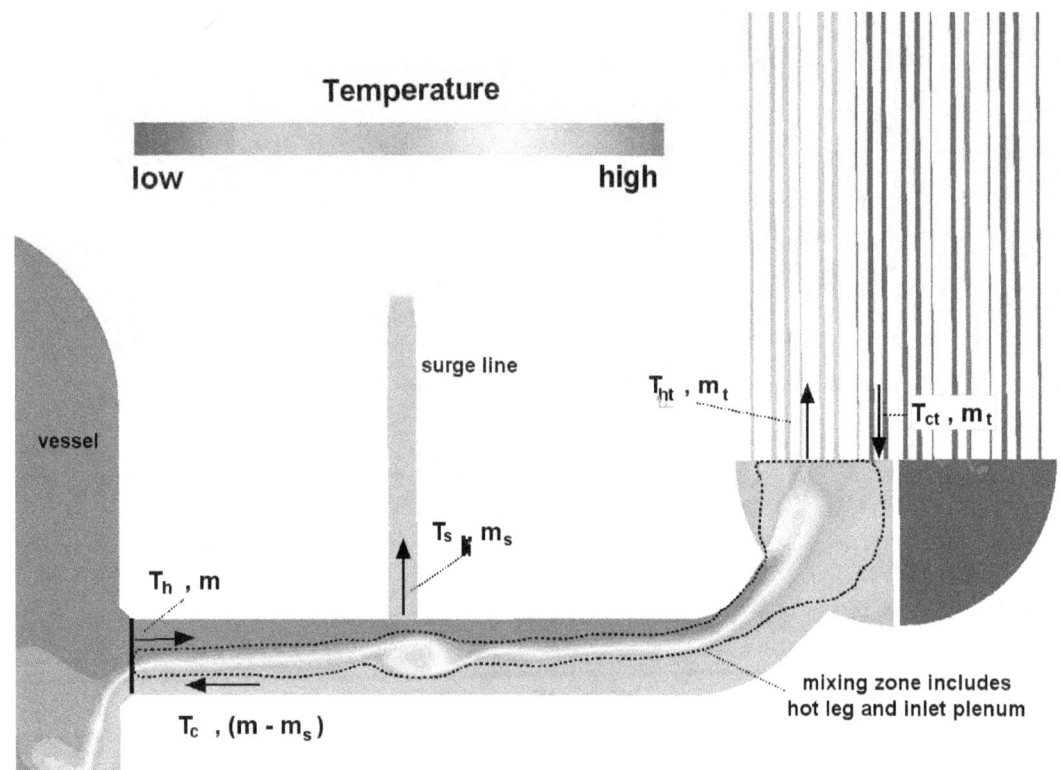

Figure 13. CFD Predictions Indicating Expanded Mixing Region

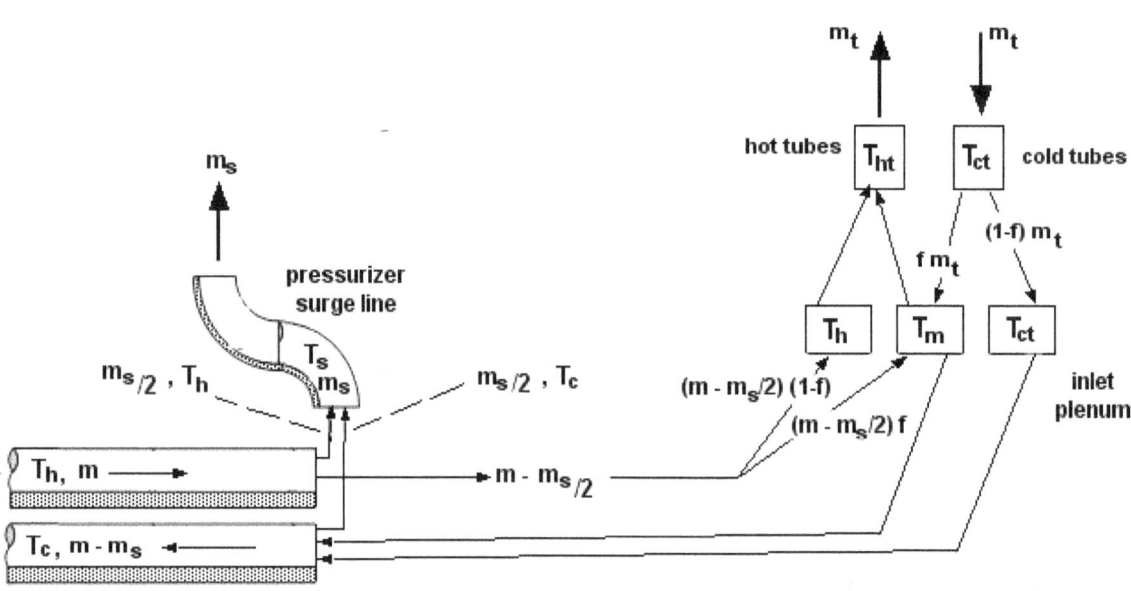

Figure 14. Mixing Model Diagram Including Hot Leg and Surge Line Flows

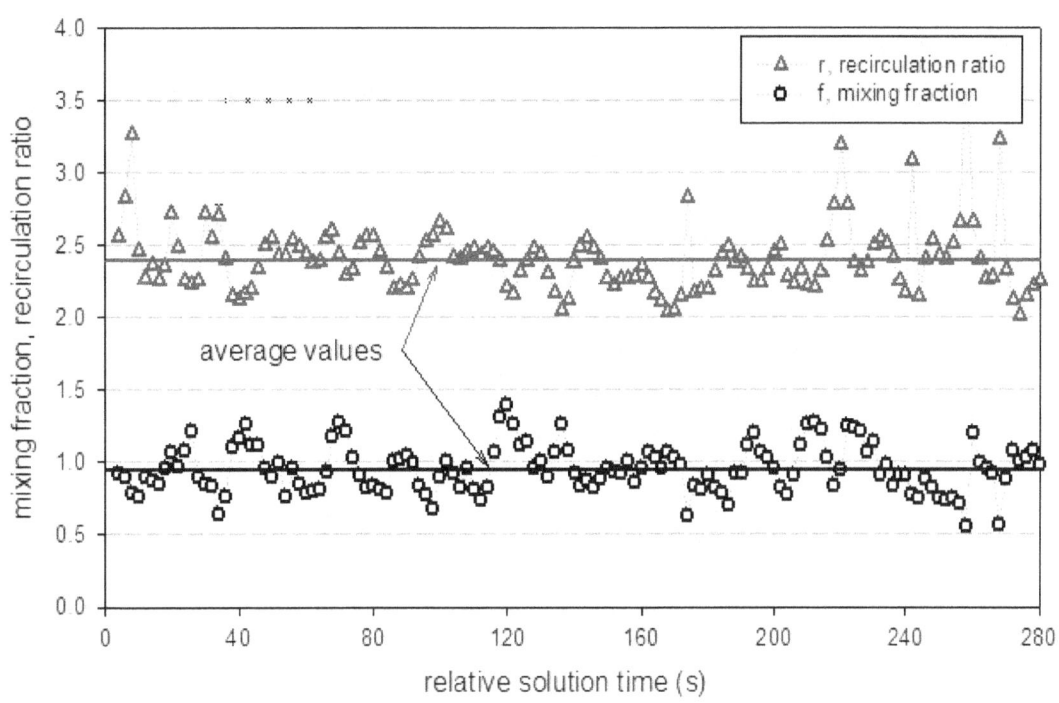

Figure 15. Mixing Fraction and Recirculation Ratio Variation with Time

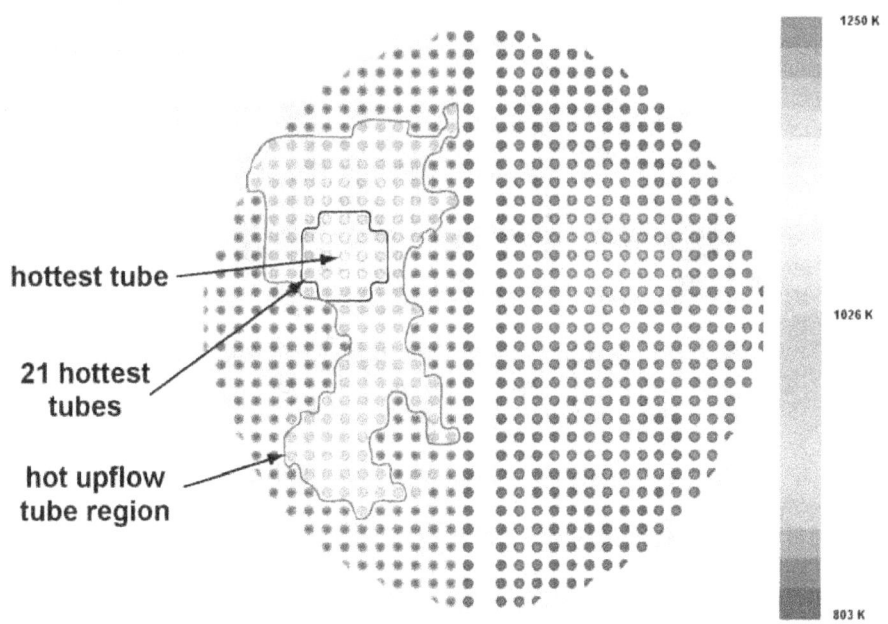

Figure 16. Temperature Contours at Tube-Sheet Entrance with Hot Tube Regions Indicated

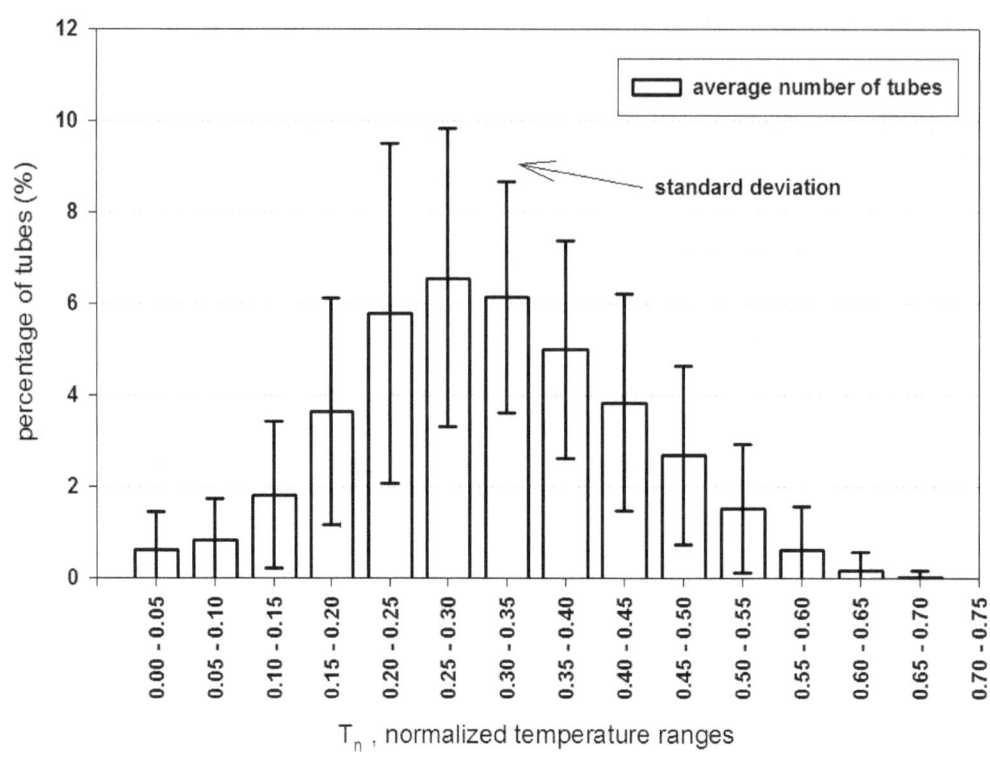

Figure 17. Range of Normalized Temperatures Observed For Hot Flow Tubes

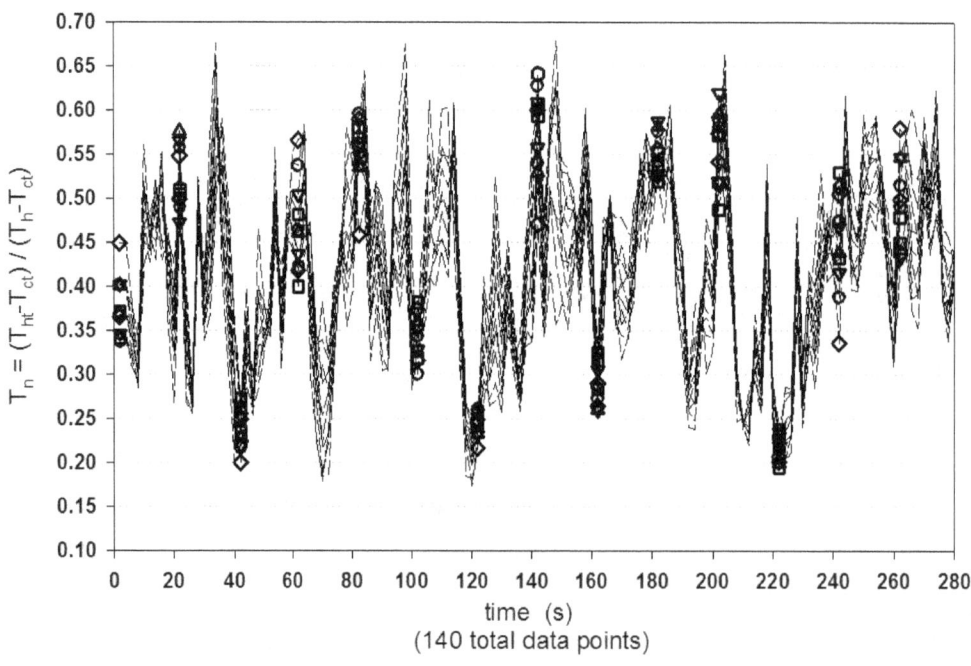

Figure 18. Normalized Temperature Variations for 10 Hottest Average Tubes

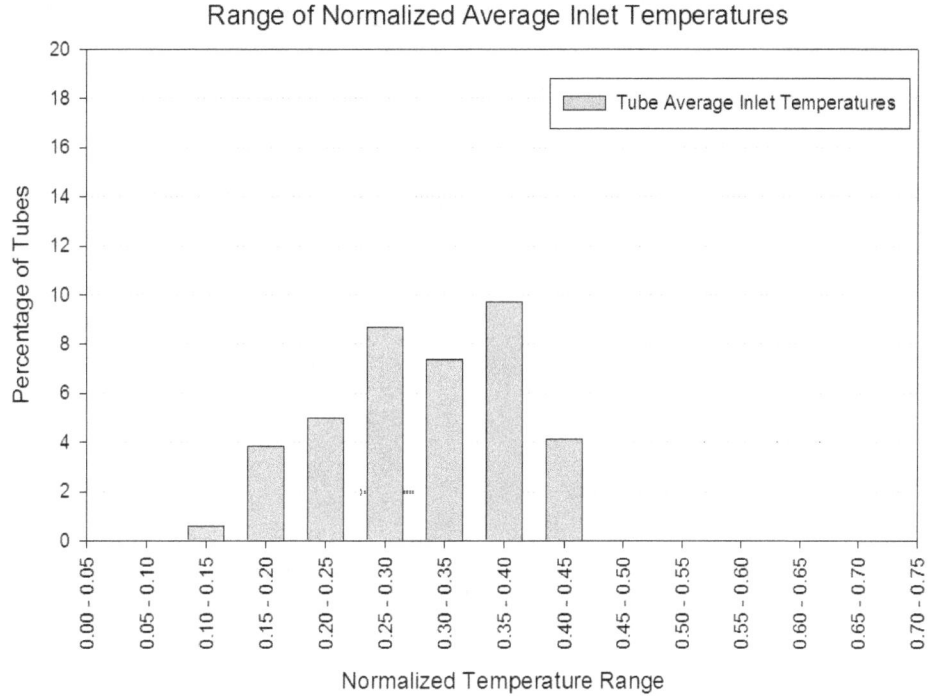

Figure 19. Average Normalized Temperatures at Tube Entrance

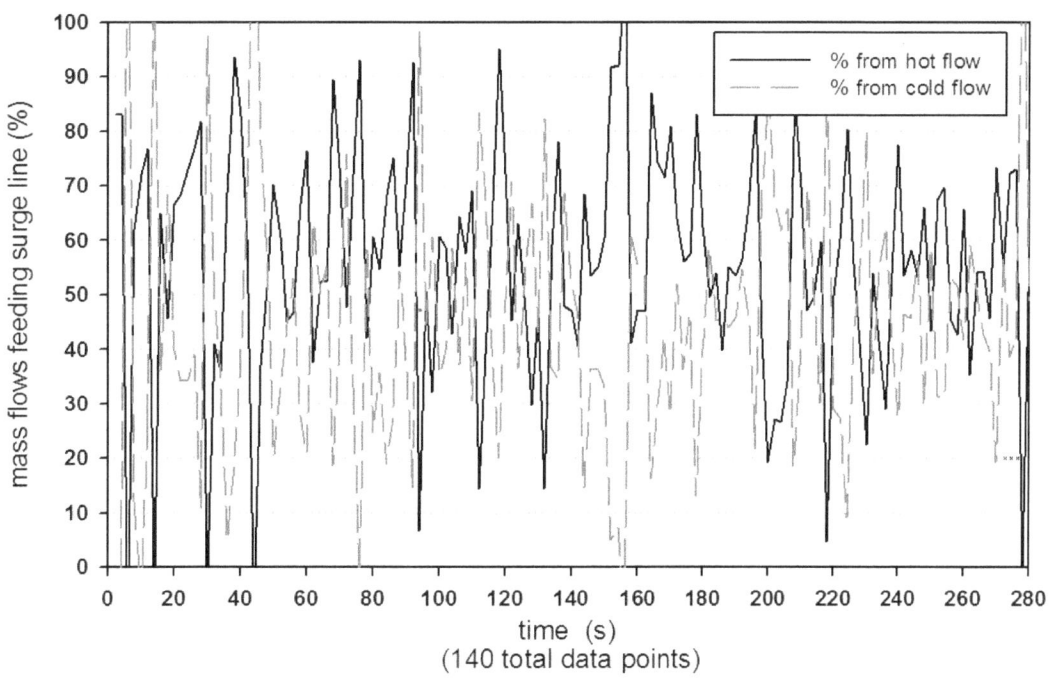

Figure 20. Mass Flows Feeding the Surge Line From Upper and Lower Hot Leg

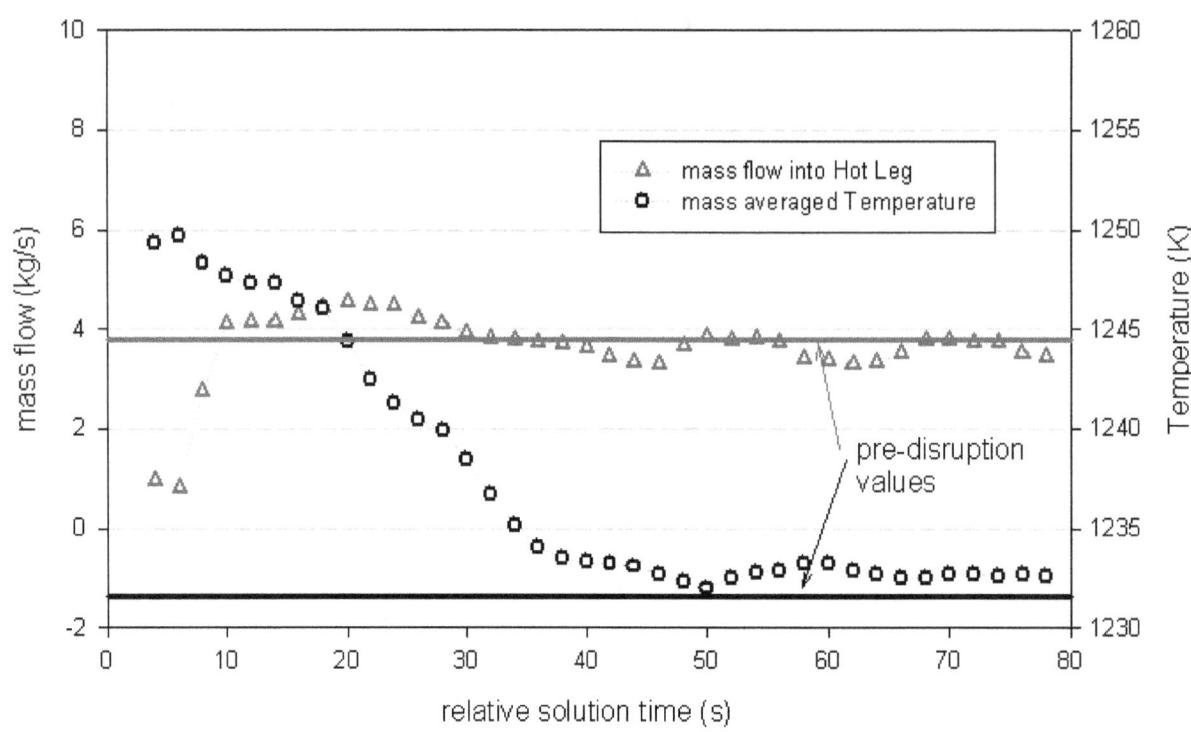

Figure 21. Reestablishment of Natural Circulation Flows in the Hot Leg after PORV is Closed

APPENDIX A

STEAM GENERATOR TUBE BUNDLE
MODELING APPROACH

A.1 BACKGROUND

The computational fluid dynamics (CFD) analysis documented in this report uses a porous media approach to model the steam generator tube bundle. Resources are not readily available to model the 3,000 or more tubes in a steam generator directly so an alternative approach is needed. The approach utilized combines groups of nine tubes (3 x 3 array) into a single equivalent-flow-area tube and incorporates models to augment the heat transfer and pressure-drop characteristics of the single tube such that it behaves like the prototypical nine-tube array that it replaces. This approach reduces the amount of computer resources required to model the tube bundle by an order of magnitude, thus making the simulation practical.

The tube-bundle model provides a flow path from the inlet to outlet plenum on the primary side of the steam generator. Flow rates through the bundle are a balance between the buoyancy driving forces and the flow resistance. To obtain the proper balance of driving force and resistance, the height, heat transfer, and pressure drop characteristics of the tube bundle must be correct. This ensures that the flow rates and temperatures in the tube bundle are predicted in a realistic manner. The modeling approach used here seeks to create a model of the tube bundle with appropriate height, heat transfer, and resistance characteristics using a model with only 1/9th of the total number of tube flow paths.

The height of the tube bundle is modeled directly and the total flow area is preserved by ensuring that each tube in the CFD model has the same flow area as the nine tubes that it replaces. This setup provides the correct pressure drop for flow entering and exiting the bundle and correctly accounts for the gravity and height aspects of the bundle. Additional modeling is needed, however, to account for the fact that the hydraulic diameter of the tubes is larger than the prototype. For a given flow rate, the larger diameter pipe in the CFD model results in pressure drops and heat-transfer rates that are lower than those expected in the actual tube bundle. The porous media formulation in the FLUENT CFD code is used to add flow resistance (pressure drop) and to enhance the heat transfer with adjustable models. These models form an important component of the tube bundle modeling approach.

Two models, available within the FLUENT modeling formulation, are utilized to adjust the behavior of the tube-bundle. One model, Equation A-1, applies a volume-based source term to the momentum equation to simulate the pressure loss through a porous region. This model is used to augment the viscous flow losses through the tubes.

$$\frac{dp}{dx} = D \cdot \mu V + C \cdot \frac{1}{2} \rho V \cdot |V| \qquad \text{A-1}$$

The first term in Equation A-1 represents viscosity-dominated loss and the second term provides a higher order term for dynamic pressure losses. In three dimensions, the coefficients C and D are matrices to allow for nonisotropic behavior.

The second model used for the tube-bundle model, Equation A-2, establishes the effective thermal conductivity, k_{eff}, of the porous media region as a volume-weighted average of the fluid and the solid conductivity values. The solid material thermal conductivity is k_s, and Φ is the void fraction. This model is used to enhance the heat transfer across the tube and effectively increases the rate of heat transfer from the tube bundle to the secondary side of the steam generator.

$$k_{eff} = \Phi \cdot k_f + (1 - \Phi)k_s \qquad\qquad\qquad\qquad \text{A-2}$$

The general approach used here is to develop a detailed model of a section of the prototypical tube bundle and to use this model to predict best-estimate pressure drop and heat transfer characteristics. Next, a corresponding section of the simplified tube bundle is modeled, and the parameters D, C, k_s, and Φ are adjusted in equations A-1 and A-2 to ensure the simplified tube bundle matches the best estimate predictions.

A.2 GEOMETRY

The prototypical tube-bundle model is loosely based on a Westinghouse model 51 steam generator with the following assumptions. There are 3,388 U-shaped tubes with a 0.019685 m (0.775 in) inner diameter and a 0.022225 m (0.875 in) outer diameter. The tubes are anchored into a 0.5398 m (21.25 in) thick tube sheet that caps off the inlet and outlet plenums. Moreover, the tubes are arranged into a 0.03254 m (1.281 in) square pitch array. The tubes are straight for a distance of 9.612 m (378.425 in) (from the tube entrance), and the total height of the tube bundle above the lower tube sheet face is 11.164 m (439.528 in).

The simplified CFD model utilizes the equivalent of 371 total tubes, each with a cross sectional area that is nine times the size of the prototype. The CFD model uses a tube pitch of 0.09761 m (3.843 in) on a square pitch array and effectively combines nine tubes from the prototype into a single tube. Figure A-1 shows a section of tubes from the prototype model compared to the equivalent tube from the simplified CFD model.

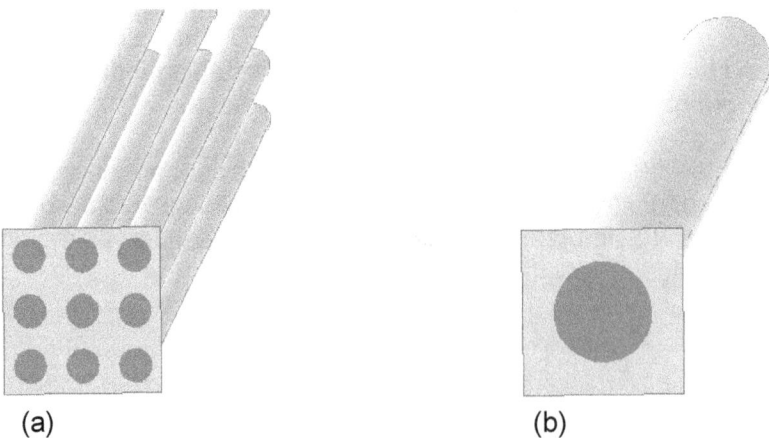

(a)   (b)

Figure A-1. (a) Nine Prototype Tubes at Tube Sheet; (b) Simplified CFD Model Used to Represent Nine Tubes.

Table 1 in the report provides a comparison of the prototype and the simplified CFD model for the steam generator.

A.3 THERMAL PROPERTIES

The heat-transfer characteristics are corrected in the simplified model by comparing the mass-averaged temperature predictions along the pipe with the best-estimate predictions of the prototype model. If the temperature drop along the pipe is consistent, then the buoyancy driving force on the tube bundles will be consistent. With no added adjustment, the heat-transfer rate from the simplified CFD model is too low and the pipe flow temperatures remain too high. Early attempts to simply increase the secondary side heat transfer rate did not reduce the pipe flow temperatures fast enough. The prototype model simply represents a much more efficient heat-transfer device compared to the simplified model. The porous media model, which relies on equation A-2 to determine the effective thermal conductivity of the fluid in the tube, is utilized as a means of increasing the heat-transfer rate across the simplified CFD model tube.

In the prototype model, the tube-boundary conditions are selected to ensure that the temperature drop along the tube is consistent with the temperature drop predicted under severe accident conditions using SCDAP/RELAP5 results as a reference. The SCDAP/RELAP5 predictions of the 1-dimensional tube flows and heat transfer are considered reasonable, and these predictions also provide our best available prediction of the conditions outside of the tubes (secondary side conditions) during the severe accident. The boundary condition above the tube sheet uses a convective boundary condition on the outer wall of the tubes and an ambient temperature (secondary side temperature) obtained directly from the SCDAP/RELAP5 predictions. Within the tube-sheet region, the tube inner wall boundary condition is set to a fixed temperature that also comes directly from the SCDAP/RELAP5 results. The prototype CFD model predictions compared very well to the SCDAP/RELAP5 predictions.

The next step is to utilize the simplified CFD model along with the porous media approach to reproduce the results of the prototype model. To use Equation A-2, a solid material needs to be defined along with a porosity value. The fluid phase thermal conductivity, k_f, is defined by the temperature-dependant fluid properties of the steam-hydrogen mixture in the reactor coolant system. The porosity, ϕ, and a solid thermal conductivity, k_s, are essentially created solely for tuning the model. The ficticous material properties for the porous solid material are outlined in Table A-1. These final solid material properties are utilized in the model, along with values for porosity and the external heat-transfer boundary condition to ensure that the simplified CFD model is consistent with the best-estimate prototype predictions. A series of trial and error studies are completed to optimize the results.

Table A-1. Fictitious Material Properties used in Model Tuning

Solid Properties	
Density (kg/m^3)	40
Cp (j/m^3 K)	2,669
k_s (W/mK)	10

A porosity value of 0.95 is found to produce good heat-transfer rates in the tubes above the tube sheet, but it did not provide enough heat transfer in the tube sheet region where the heat transfer is predicted to be the highest in the best-estimate model. A porosity of 0.7 is used in the tube sheet region. The simplified CFD also uses a fixed wall temperature (same value) for the inner wall of the tube in the tube-sheet region. Above the tube sheet, a convective boundary condition is used on the outer surface of the tubes with the same ambient temperature as the best-estimate model. The heat-transfer coefficient is increased to 86 W/m2-K and was the final parameter adjusted to ensure consistency between the results.

Figure A-2 shows the predicted temperatures along the tube flow path for both models. The best-estimate prototype prediction for three different mass-average tube velocities is shown with the dotted lines. The prediction at 0.15 m/s matches the SCDAP/RELAP5 predictions which also had tube velocities of 0.15 m/s. The simplified tube model is shown with the symbols. At 0.15 m/s, the simplified model matches the best estimate predictions which also match the temperature drop in the SCDAP/RELAP5 predictions. Two variations are considered since the CFD model will have variations in the tube velocities. For each velocity boundary condition, the simplified model predictions match the best estimate predictions very well. The final parameters used in the simplified model are passed along to the full CFD model of the steam generator tube bundle. This approach ensures that the simplified model is consistent with the tube bundle heat transfer rates from SCDAP/REALP5.

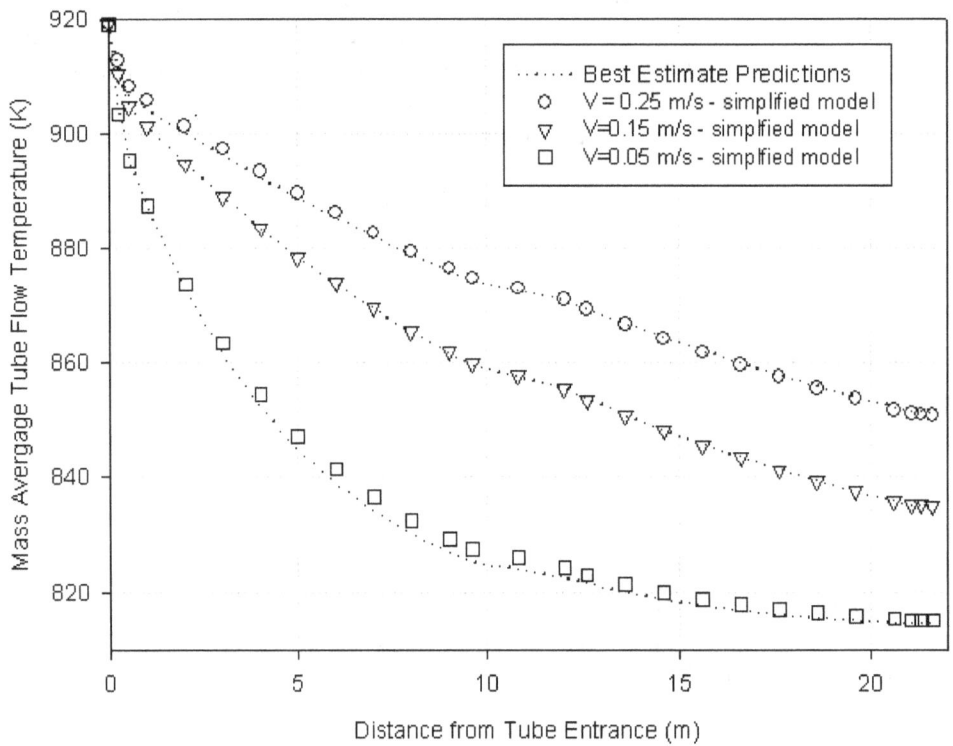

Figure A-2. Gas Temperature vs. Flow Distance in the Tube

A.4 LOSS COEFFICIENT DETERMINATION

The loss coefficients needed in Equation A-1 are determined by comparing the pressure drop along the tubes between the best-estimate and simplified models. Both models predicted the same entrance losses for flow entering the tube sheet from the inlet plenum. No additional modeling is needed for entrance loss predictions. Predictions for the viscous sheer loss along the tube, dp/dx, are too low in the base simplified CFD model due to the reduced wetted surface area. This is corrected by using Equation A-1 to augment the pressure drop. Using values in Equation A-1 for C and D of 122000, and 0.6422, respectively, the simplified tube model is adjusted to match the best estimate pressure drop predictions over a wide velocity range. Figure A-3 illustrates the predicted results for the pressure along the tube flow path for three different mass averaged velocity cases. The agreement is good in all three cases.

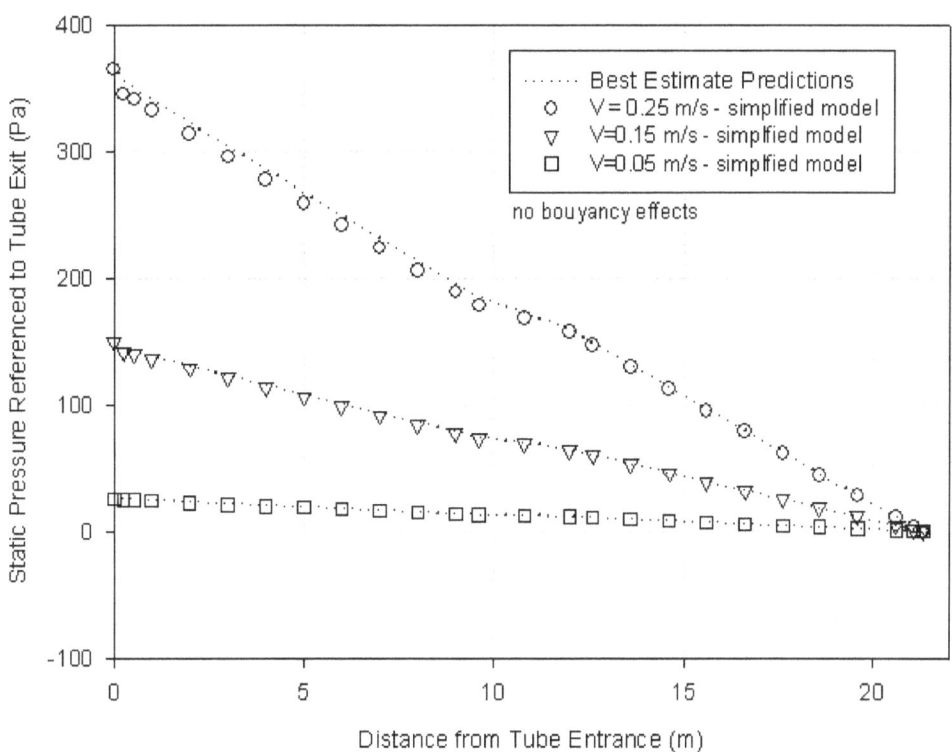

Figure A-3. Pressure vs. Flow Distance in the Tube

A.5 SUMMARY

A simplified tube model, utilizing the porous media formulation in FLUENT to enhance flow resistance and heat transfer, is demonstrated to yield similar flow resistance and heat-transfer predictions as a more refined best-estimate model. The noding assumptions, tube size, and all modeling assumptions used for this demonstration are applied in a full scale tube-bundle model. The simplified tube bundle model allows the model to simulate a tube bundle with thousands of tubes using only hundreds of tube flow paths.

NRC FORM 335 (9-2004) NRCMD 3.7	U.S. NUCLEAR REGULATORY COMMISSION	1. REPORT NUMBER (Assigned by NRC, Add Vol., Supp., Rev., and Addendum Numbers, if any.)
	BIBLIOGRAPHIC DATA SHEET *(See instructions on the reverse)*	NUREG-1922

2. TITLE AND SUBTITLE

Computational Fluid Dynamics Analysis of Natural Circulation Flows in a Pressurized-Water Reactor Loop under Severe Accident Conditions

3. DATE REPORT PUBLISHED

MONTH	YEAR
February	2010

4. FIN OR GRANT NUMBER

5. AUTHOR(S)

Christopher F. Boyd
Kenneth W. Armstrong

6. TYPE OF REPORT

Technical

7. PERIOD COVERED *(Inclusive Dates)*

8. PERFORMING ORGANIZATION - NAME AND ADDRESS *(If NRC, provide Division, Office or Region, U.S. Nuclear Regulatory Commission, and mailing address; if contractor, provide name and mailing address.)*

Division of System Analysis
Office of Nuclear Regulatory Research
U.S. Nuclear Regulatory Commission
Washington, DC 20555-0001

9. SPONSORING ORGANIZATION - NAME AND ADDRESS *(If NRC, type "Same as above"; if contractor, provide NRC Division, Office or Region, U.S. Nuclear Regulatory Commission, and mailing address.)*

See above

10. SUPPLEMENTARY NOTES

11. ABSTRACT *(200 words or less)*

Computational fluid dynamics is used to predict the natural circulation flows between a simplified reactor vessel and the steam generator of a pressurized-water reactor (PWR) during a severe accident scenario. The results extend earlier predictions of steam generator inlet plenum mixing with the inclusion of the entire natural circulation loop between the reactor vessel upper plenum and the steam generator. Tube leakage and mass flow into the pressurizer surge line are also considered. The predictions are utilized as a numerical experiment to improve the basis for simplified models applied in one-dimensional system codes that are used during the prediction of severe accident natural circulation flows. An updated inlet plenum mixing model is proposed that accounts for mixing in the hot leg too. The new model is consistent with the predicted behavior and accounts for flow into a side mounted surge line if present. A density-based Froude number correlation is utilized to provide a method for determining the flow rate from the vessel to the hot leg directly from the conditions at the ends of the hot leg pipe. This provides a physically based approach for establishing the hot leg flows. The mixing parameters and correlations are proposed as a best-estimate approach for estimating the flow rates and mixing in one-dimensional system codes applied to severe accident natural circulation conditions. Sensitivity studies demonstrate the applicability of the approach over a range of conditions. The predictions are most sensitive to changes in the steam generator secondary side temperatures or heat transfer rates to the steam generator. Grid independence is demonstrated through comparisons with previous models and by increasing the number of cells in the model. A further modeling improvement is suggested regarding the application of thermal entrance effects in the hot leg and surge line. This work supports the U.S. Nuclear Regulatory Commission studies of steam generator tube integrity under severe accident conditions.

12. KEY WORDS/DESCRIPTORS *(List words or phrases that will assist researchers in locating the report.)*

computational fluid dynamics
induced failure
natural circulation flow
severe accident
steam generator inlet plenum mixing
tube leakage

13. AVAILABILITY STATEMENT

unlimited

14. SECURITY CLASS FICATION

(This Page)

unclassified

(This Report)

unclassified

15. NUMBER OF PAGES

16. PRICE

PR NTED ON RECYCLED PAPER

www.ingramcontent.com/pod-product-compliance
Lightning Source LLC
Chambersburg PA
CBHW081600170526
45166CB00009B/2771